TRANSACTIONS
of the
American Philosophical Society
Held at Philadelphia for Promoting Useful Knowledge

VOLUME 76, Part 5, 1986

Medieval Agriculture, the Southern French Countryside, and the Early Cistercians.
A Study of Forty-three Monasteries

CONSTANCE HOFFMAN BERMAN
Washington, D.C.

THE AMERICAN PHILOSOPHICAL SOCIETY

Independence Square, Philadelphia

Reprinted 1992.

Copyright © 1986 by The American Philosophical Society

Library of Congress Catalog
Card Number 84-71079
International Standard Book Number 0-87169-765-3
US ISSN 0065-9746

To the memory of
my maternal grandfather,
William B. Stevenson,
a farmer and builder,

and to David and Benjamin.

"Generatio preterit et generatio advenit, terra autem in eternum stat."
　　　Charter of Humbert of Savoy to Bonnevaux, ca. 1165.

CONTENTS

List of Maps and Tables	viii
Preface	ix
I. Foundations of the Abbeys and the Origins of the Traditional Picture of the Cistercian Economy	1
II. The Question of New Lands and Reclamation by Cistercians	11
III. Consolidation of Rights and Recruiting Personnel	31
IV. The Profits of Grange Agriculture	61
V. Pastoralism and Transhumance	94
VI. Cistercians and Regional Economic Growth	118
Maps	130
Tables	135
Appendix I	
Assessments of Southern French Cistercian Abbeys by Filiation	142
Appendix II	
Distribution of Profits from Agriculture	144
Sources Consulted and Bibliography	148
Index	165

LIST OF MAPS

Map 1: Cistercian Monasteries in Southern France 130
Map 2: Cistercian Granges near Toulouse 131
Map 3: Cistercian Granges in Languedoc 132
Map 4: Cistercian Granges and Pastoralism in the Rouergue 133
Map 5: General Directions of Cistercian Transhumance 134

LIST OF TABLES

Table 1: Cistercian Foundations and Incorporations 135
Table 2: Surviving Documents and Total Cash Expenditures 137
Table 3: Number of Documented Granges for Cistercian Houses 141

PREFACE

This is a study of medieval agriculture, of the rural world of southern France, and of the early corporate farms of the new religious order of Cîteaux, founded in Burgundy in 1098 and imported into southern France in the mid-twelfth century. It is a study of the agriculture and pastoralism practiced by the white monks, as the Cistercians were often called, in a region which is both vast and varied in topography, climate, and custom. It also assesses that order's contributions to southern-French economic development in the twelfth and thirteenth centuries. It shows that the Cistercians in that region did not acquire lands for their huge, newly consolidated farms—the granges—through clearance and reclamation of unoccupied lands as traditional accounts suggest, but rather through the careful purchase and reorganization of holdings which had often had a long history of cultivation.

When this research was begun, in 1972, the contention that Cistercians did not create new lands for themselves out of previously uncultivated wilderness was probably the most startling and important part of this assessment. At the time that this goes to press, however, the conclusion of chapters 1 and 2, that Cistercians were not the "pioneers" which they were once thought, has been confirmed by studies of Cistercians in other parts of Europe.[1] This recognition by specialists in the field of monastic history and geography, changes the historian's task. It is now necessary not only to show that the older models of Cistercian economic practice are incorrect, but to describe how the order's abbeys accomplished their tremendous economic successes. In the third chapter of this study the managerial process by which land was purchased, tithes acquired, and laborers recruited is discussed. This is a description of how, if not by clearance, Cistercians nonetheless developed a patrimony; the links described between land acquisition, recruitment of *conversi* or lay-brothers, and tithe acquisition are crucial to the explanation of the order's early growth and prosperity. Perhaps most interesting to the student of rural economy are chapters 4 and 5 which describe respectively the workings and profitability of the Cistercian

---

[1] See the updated versions of papers presented at the conference on the Cistercian grange held at the abbey of Flaran in Gascony in 1981, by Robert Fossier, "L'économie cistercienne dans les plaines du nord-ouest de l'Europe," *L'économie cistercienne. Géographie—Mutations du Moyen Age aux Temps modernes. Publications de la Commission d'histoire de Flaran*, 3 (Auch: Comité départemental du Tourisme du Gers, 1983), 53–74, and Charles Higounet, "Essai sur les granges cisterciennes," *L'économie cistercienne. Géographie—Mutations du Moyen Age aux Temps modernes. Publications de la Commission d'histoire de Flaran*, 3 (Auch: Comité départemental du Tourisme du Gers, 1983), 157–80.

grange—the corporate farm *par excellence* of the High Middle Ages, and the institution of intensive pastoralism and transhumance by the order's abbeys. These two aspects of Cistercian economic practice are reconstructed from the numerous but extremely unwieldy land acquisition documents which survive. It has also been possible to suggest an increased profitability or at least a relative increase in profitability of Cistercian cereal cultivation and animal husbandry despite the lack of early production records for these farms. The picture presented of Cistercian agriculture and pastoralism adds to our overall understanding of medieval rural life and of the improvements in that life caused by the introduction of technology and rationalization. This reconstruction of how Cistercians practiced pastoralism and agriculture, as well as information on land acquisition in chapter 3 and on urban markets and interests in chapter 6, show how the poor Cistercian monks of the mid-twelfth century in southern France became the wealthy Cistercians of the thirteenth.

In describing the order's rationalization and management of agriculture, this book assesses the profitability of land cultivation by a new class of laborers, the *conversi* or lay-brothers. It also describes the order's assiduous collection of pasture rights which enabled its abbeys to improve the ratio of animals per acre under cereal cultivation through the introduction of seasonal pastoral migration, or transhumance. Cistercian stock-raising, during a period of rising demand for meat, leather, parchment, and other animal products, provided an important source of revenues to the new order, one which funded further land acquisitions. Market privileges, concessions allowing the monks to travel and move their goods without tolls, and tithe exemptions also improved profits when surpluses were sold in the growing neighboring towns.

This book covers a region in which forty-three Cistercian houses for men were founded in the twelfth and thirteenth centuries. Only the houses for men are considered; from the point of view of economic practice, houses for women were much more diverse. Moreover, the precise relationship of women to the Cistercian order in the twelfth century is debatable, requiring more attention than can be devoted to it here.[2] This study is based on a wealth of archival material, but because records of different houses survive to various degrees, it concentrates on those abbeys which kept good records in the twelfth and thirteenth centuries and which were fortunate enough to have had those records survive. This causes an initial and unavoidable bias toward the more successful houses of the region since wealthier houses could afford to preserve more records. Conclusions rest primarily on records for the houses of Berdoues and Gimont in Gascony; Grandselve and Bel-

---

[2] See Sally Thompson, "The Problem of the Cistercian Nuns in the Twelfth and early Thirteenth Centuries," in *Medieval Women*, ed. Derek Baker, *Studies in Church History: Subsidia* 1, (Oxford: Blackwell, 1978), 227–52.

PREFACE

leperche north of Toulouse; Boulbonne, and Calers in the Ariège south of Toulouse; Candeil in the Albigeois, and Villelongue near Carcassonne; Fontfroide, Valmagne, and Franquevaux on the Languedoc coast; Bonnecombe, Bonneval, and Silvanès in the Rouergue; and Bonnevaux and Léoncel in the marquisate of Provence, east of the Rhône.

Although some materials are in published collections, the major evidence comes from unpublished documents. Not counting the many duplicates of individual acts, over seven thousand individual contracts have survived and were collected for this study. Most of these contracts, expressed in the form of private charters, are in copies made by the monks themselves in the twelfth and thirteenth centuries, which when transcribed into codices made up what are called monastic cartularies. In addition, there are some originals available, or copies contemporary with originals which survive as loose parchments in the archives, as well as later copies made for use in legal disputes, copies made by various religious and royal commissions (most notably the Doat Commission), and a variety of inventories of archives compiled for the monks, dating to as early as the fifteenth century, some of which have nearly complete transcriptions of each act. Using such extensive materials, all of which have been consulted in an effort to make this an exhaustive study of the Cistercians in southern France, it is possible to make generalizations about the order in that region. It has not been possible, however, to discuss every property held by the Cistercians in southern France, or even to discuss all aspects of an abbey's economy in detail, although an attempt has been made to show that the material cited is representative. Many institutional issues such as dating of foundations, lists of early abbots, etc., are considered elsewhere.[3]

Any examination of Cistercian economic practice draws on a well-established tradition of local studies of early Cistercian houses, which have increased considerably in the decades since 1953, the eighth centenary of Bernard of Clairvaux's death.[4] Although many of the individual practices mentioned here have been described by historians of Cistercian practice in other parts of Europe, in the interests of space and clarity explicit comparisons are kept to a minimum or left to the reader. It is certain, however, that without such earlier work, the conclusions here about how the various aspects of Cistercian economic practice in southern France relate to one another would have been much less complete. An attempt is made to con-

---

[3] Constance H. Berman, "The Cistercians in the County of Toulouse: 1132–1249, the Order's Foundations and Land Acquisitions," Ph. D. dissertation, University of Wisconsin, 1978. Details of foundation history are discussed in "The Growth of the Cistercian Order in Southern France," *Analecta cisterciensia*, forthcoming.

[4] Louis J. Lekai, *The Cistercians, Ideal and Reality* (Kent, Ohio: Kent State University Press, 1977); R. A. Donkin, "A Checklist of Printed Works relating to the Cistercian Order," *Documentation cistercienne* 2 (Rochefort, Belgium: Abbaye N. D. de St. Rémy, 1969); Giles Constable, *Medieval Monasticism. A Select Bibliography* (Toronto: University of Toronto Press, 1976), and bibliography below.

sider the history of Cistercian economic practice outside the realm of "ideal" versus "reality," Golden Age and decadence, and other schemes used by historians to contrast early Cistercian legislation with local practice. Cistercian practice in southern France from the very outset diverged from the ideals of the order's legislators—necessarily so, because of local conditions. Indeed, most monastic historians now agree that Cistercian foundations and settlement patterns in many parts of Europe were different from what would be expected from a reading of the order's legislation alone.[5] Because despite recent studies, however, earlier theories about the contribution of Cistercians to twelfth-century economic growth have become so interwoven with descriptions of the order based on legislative ideals, it is important to show that Cistercian activities in southern France do not fit the old economic model. The primary focus on the order's economic practice in the southern French countryside underlines the ties between various aspects of Cistercian agricultural practice as well as the monks' dependence on conditions in the region into which they came. This description of the economic practices of a specific group in a particular region should concern historians not only of religious life, monastic economy, and the Cistercian order, but also of medieval settlement, demographic growth, and economic development.

This work began as a doctoral dissertation at the University of Wisconsin, but it has grown far beyond that. It has received the encouragement of too many individuals to list them all here, although a few must be formally thanked. David Herlihy introduced me to a Cistercian cartulary, on which I wrote my master's thesis. Barbara Kreutz, Maureen Mazzaoui, and William Courtenay supported the completion of my dissertation. Fr. Louis J. Lekai, E. Rozanne Elder, and others associated with the Cistercian Studies Institute in Kalamazoo have offered enormous aid and encouragement. I would also like to acknowledge help at critical moments from Richard Unger, Giles Constable, Georges Duby, Wendy Pfeffer, R. A. Donkin, and Carole Le Faivre. Maps were prepared by Irene Jewett. Fr. Edmund Mikkers, editor of *Cîteaux*, has kindly granted me permission to cite material from an article published in that journal, and Kendall and Hunt Publishing Company has permitted me to reproduce the frontispiece photo used here; it originally appeared in my article "Fortified Monastic Granges in the Rouergue," in *The Medieval Castle: Romance and Reality*, ed. Kathryn Reyerson and Faye Powe (Dubuque, Iowa: 1985). None of them or the many others not mentioned here should be considered responsible for my conclusions.

A number of institutions have aided my research. Funds were provided by the University of Wisconsin graduate school and history department, the Vilas Foundation, the American Association of University Women—Madison Branch, the American Philosophical Society's Penrose Fund, and an American Council of Learned Societies Grant-in-Aid. The Library of

---

[5] See below, especially p. 33 n. 20.

PREFACE

Congress has provided me with work space while completing the writing, the University of Wisconsin and the University of British Columbia cooperated in supplying me with published materials through Interlibrary Loan while I lived in Canada, and the Scandinavian Interlibrary Loan Services did the same during a year in Trondheim, Norway; Bard College gave work-study support.

Although a certain number of the documents consulted were in published editions, in a typescript edition of a Grandselve cartulary generously lent me by Mme Mireille Castaing-Sicard, or on microfilm, most were archival documents consulted in the Bibliothèque nationale and Archives nationales in Paris and in municipal and departmental archives and libraries or other collections in the region of the study. Several private learned societies, most notably the Société des lettres, sciences, et arts de l'Aveyron in Rodez, to which I was graciously elected to membership, also provided me access to their documents. Working in Toulouse, Rodez, Narbonne, Poitiers, Montpellier, Nîmes, Albi, and Carcassonne was not only a necessity but a pleasure, and I have many fond memories of helpful archivists and the members of those archives' staff, in particular Mlle Denise Rouanet, Municipal Librarian of Narbonne, who granted me permission to make a microfilm for my own use. Very important were the opportunities to explore the countryside, meet the local peasants, photograph the surviving granges, eat the magnificent foods of the region, and travel the back roads, sometimes on foot. Such ventures gave me insights into the country life and the scale of things which no books or maps could possibly convey. Those trips gained me many friends and taught me to speak in the accents of the Midi. They made me love the region on which I was working, and made me want to return there often.

During all these peregrinations, my note cards and lists, and manuscript drafts have found their way several times from one side of the Atlantic to the other or from one side of the North American continent and back. As I look back at the history of this book, I realize how much tedious and sometimes back-breaking labor, as well as a number of perceptive comments, have been given it by my husband David. Although the study is dedicated first to a grandparent who gave me my first experience of cows and meadows and barns, it was David who was dedicated enough to my professional goals to encourage me to complete it. I cannot possibly thank him enough for his support, or Benjamin enough for his smiles.

Washington, D.C.
June 1986

CONSTANCE HOFFMAN BERMAN

A NOTE ON THE DOCUMENTATION AND DATES

About half of the documents consulted for this study are in published collections. Published document references are cited using the numbering systems established in the various texts, followed by a date (i.e., *Cart. de Berdoues*, no. 125 [1137], or *Rec. de Bonnefont*, no. 13 [n.d.]). Unpublished documents are generally cited by the numbering system found in the cartulary or other text or by numbering from beginning to end rather than by folio number. References to unpublished documents are easily distinguished from published documents since they are not italicized (e.g., Rodez, Soc. des lettres, Bonnecombe, "Cart. d'Iz et Bougaunes," no. 33 [1185], or Carcassonne, A. D. Aude, "Inv. de Fonfroide," fol. 36 [1057]). Since many documents lapse into the vernacular or show peculiarities of Latin style, I have normalized the Latin cited and inserted the editorial *sic* only in cases where it is necessary for clarity. Dates are generally given as read, without conversion into new-style calendar dates, since many are recorded without giving the exact month or day. In a few cases, this may cause a discrepancy of not more than one year from published editions. Also, please note that references to the Bonnevaux Cartulary all refer to the Chevalier volume.

I. FOUNDATIONS OF THE ABBEYS AND THE ORIGINS OF THE TRADITIONAL PICTURE OF THE CISTERCIAN ECONOMY

A discussion of traditional notions of Cistercian economic practice must begin with the founding of the order itself and its early legislation. A monastic foundation was made at Cîteaux in 1098; this semi-eremitical settlement in Burgundy nearly did not survive. Fortunately, Cîteaux eventually attracted recruits, among them the enthusiastic Bernard of Fontaines (1090–1153) who became the famous abbot of Clairvaux. After an initial period of organization, daughter-houses began to be founded on the model of Cîteaux, and they in turn established their own daughter-houses, farther and farther from Burgundy, until a huge order of new, reformed monks had spread over western Europe and beyond. Although claiming only to be restoring the early purity of St. Benedict's unembellished rule, the Cistercians introduced certain new "customs" into their monastic practice. In the economic sphere these included a search for simple, frugal, and ascetic lives, an attempt to divorce Cistercian land cultivation from manorialism by the introduction of direct cultivation on "granges" by lay brothers, and an effort to find solitude and keep the world at arm's length, which led to the idealization of their monastic foundations as having been made "far from inhabited places." These features of the new monastic institution, espoused in the early legislation of the order's General Chapter, have often been used to explain the economic success of the early Cistercians.

It was during the lifetime of Bernard of Clairvaux that the Cistercians became popular and the order began to gather wealth. The new order had profited considerably from the wave of religious reform in western Europe dating from the late eleventh century, and many of its houses were incorporations of foundations made by earlier reformers. Cistercian monks soon became powerful in the politics of Christendom and the order received important papal privileges of immunity from local episcopal jurisdiction and exemption from the payment of ecclesiastical tithes. It was partly as a result of these ecclesiastical privileges that scores of unaffiliated monasteries and hermitages began to be attracted into the new Cistercian order.[1] In

---

[1] See Lekai, *Cistercians*, passim; David Knowles, *The Monastic Order in England, 940–1216*, 2nd ed. (Cambridge: Cambridge University Press, 1963), 119–223; J.-B. Mahn, *L'ordre cistercien et son gouvernement des origines au milieu du XIIIe siècle (1098–1265)* (Paris: Boccard, 1945), parts I and II, and pp. 73–155; and J.-M. Canivez, "Cîteaux," *Dictionnaire d'histoire et du géographie ecclésiastique* 12 (1953), 852–997.

southern France, such incorporations and new Cistercian foundations were widespread, and extensive archives survive to document the history of a number of those houses in what is today the French Midi.

Houses for which considerable numbers of documents survive were spread throughout the southern French landscape in a variety of settings. Set in the open, rolling countryside of Gascony west of the Garonne, the early daughter-house of Morimond, Berdoues (founded ca. 1142), and Berdoues's own daughter-house of Gimont (founded ca. 1147), both have surviving cartularies. Although the abundant records for Gimont end abruptly in the early thirteenth century, Berdoues's charters continued to be recorded in a volume which includes acts up to the mid-thirteenth century.[2] Similarly, Grandselve, whose granges extended on both sides of the Garonne River north of Toulouse, has about fifteen hundred surviving charters for this period. In contrast, for Grandselve's neighboring abbey, Belleperche, relatively few documents survive, although Belleperche was also a powerful medieval abbey.[3] Both Grandselve and Belleperche were probably founded by Gerald of Salles, a slightly younger colleague of the western French reformer Robert of Arbrissel (ca. 1060–1117), and both Grandselve and Belleperche were only incorporated into the Cistercian order as daughters of Bernard's Clairvaux in the 1140s.[4] Almost no documents survive for abbeys in the vicinity of Montauban. For houses in the region south of Toulouse relatively few documents remain for the abbeys of Boulbonne (founded ca. 1160), Calers (founded 1148), and Bonnefont (founded 1139), in the filiations of Morimond, Clairvaux, and Morimond, respectively; nonetheless, these three abbeys are represented in certain crucial documents.[5]

For the region east of Toulouse, there is considerable evidence from three of the five Cistercian abbeys found in the Rouergue, as well as one of the two Cistercian houses of the Albigeois. Of these, Candeil (founded

---

[2] *Cartulaire de l'abbaye de Berdoues près Mirande*, ed. l'abbé Cazaurin (The Hague: Nijhoff, 1905), and *Cartulaire de l'abbaye de Gimont*, ed. A. Clergeac (Paris: Champion, 1905).

[3] There are five manscript cartularies for Grandselve, with hundreds of charters in each: Paris, Bibliothèque nationale, Latin MSS 9994, 11008, 11009, 11010, and 11011; in addition see Paris, Archives nationales, L1009 bis, and Toulouse, A.D. Haute-Garonne, 108H series. For Belleperche, see Paris, B. N. Coll. Doat, vols. 91–92. Interesting documents concerning Belleperche are also sometimes found among the Grandselve documents. Generally, Belleperche's documents, like those of la Gard-Dieu, St. Marcel, and Belloc, were destroyed in the Wars of Religion.

[4] Table 1 gives foundation and affiliation dates; on the foundation of houses by pre-Cistercian reformers in this region and their affiliation by Bernard of Clairvaux, see Cicely d'A. Angleton, "Two Cistercian Preaching Missions to Languedoc in the Twelfth Century: 1145, 1178," unpublished Ph.D. diss., Catholic University of America, 1984, and Berman, "Growth," forthcoming.

[5] An inventory of Boulbonne's documents is found in *Histoire générale de Languedoc*, ed. Cl. Devic and J. Vaissete (Toulouse: Privat, 1872–93), 8; cols. 1883–926; most surviving documents are found in Paris, B. N. Coll. Doat, vols. 83–86. For Bonnefont, there is a very recent collection of charters and charter references: *Recueil des actes de l'abbaye cistercienne de Bonnefont-en-Comminges*, ed. Ch. Samaran and Ch. Higounet (Paris: Bibliothèque nationale, 1970). On Calers, see Casimir Barrière-Flavy, *L'abbaye de Calers* (Toulouse: Chauvin, 1887–89), and A. D. Haute-Garonne, 108H Calers, liasses 4–20.

1150–1152), was sited south of Albi. A daughter-house of Grandselve, it benefited especially from gifts from the bishops of Albi. Its holdings, described in copies of documents made by the Doat Commission, extended from south of the Tarn to the Cévennes.[6] Candeil was mother-abbey of the house of Bonnecombe (founded ca. 1166) in the Rouergue. Bonnecombe's various granges and properties occupied a variety of soil and topography in the region south and west of Rodez, and are documented extensively in the almost totally surviving archives.[7] Bonnecombe's acquisitions eventually extended southward until they came into conflict with those of the smaller Cistercian abbey of Silvanès (incorporated 1138) located in the Cévennes in the southern Rouergue. Despite the abbey's small size, Silvanès's cartulary is an important source, for it contains hundreds of mid-twelfth-century documents, and the abbey has the only authentic twelfth-century chronicle. Unfortunately, few charters for the years after 1170 survive.[8] Archival material is also scarce on Bonneval (founded 1147), located on the edge of the Aubrac plateau of the northern Rouergue, although this house was as wealthy as Bonnecombe in the Middle Ages. A fire in Bonneval's archives in 1719 destroyed all but a few of its extensive twelfth- and thirteenth-century land titles. In addition, Bonneval's early foundation narratives are patently false (and were considered so in the Middle Ages).[9]

Two wealthy houses whose churches still stand on the Languedoc plain west of the Rhône were those of Fontfroide (incorporated in 1145) and

---

[6] Most of Candeil's archives are found in Paris, B. N. Coll. Doat, vols. 114–15; for the gifts by bishops of Albi, see ibid., no. 905 (1207), and Albi, A. D. Tarn, H38 "Répertoire de Candeil (1739–41)," no. 95 (1212).

[7] Documents on Bonnecombe include *Cartulaire de l'abbaye de Bonnecombe*, ed. P.-A. Verlaguet (Rodez: Carrère, 1918–25), vol. 1 (only volume ever published), plus four unpublished cartularies: Rodez, Société des lettres, sciences, et arts de l'Aveyron, MSS 1, 2, and 3, Bonnecombe, "Cartulaire de Magrin," "Cartulaire de Moncan," and "Cartulaire d'Iz et Bougaunes," and Rodez, A. D. Aveyron, H series non coté, "Cartulaire de Bernac." See also Charles Higounet, "Les types d'exploitations cisterciennes et prémontrées du XIIIe siècle et leur rôle dans la formation de l'habitat et des paysages ruraux," *Paysages et villages neufs du moyen âge. Recueil d'articles de Charles Higounet* (Bordeaux: Fédération historique du sud-ouest, 1975), 177–83, and Constance H. Berman, "Administrative Records for the Cistercian Grange: the Evidence of the Cartularies of Bonnecombe," *Cîteaux: comm. cist.* 30 (1979): 201–20.

[8] *Cartulaire de l'abbaye de Silvanès*, ed. P.-A. Verlaguet (Rodez: Carrère, 1910). Verlaguet's edition includes materials found in local notarial registers as well as the documents from Paris, B. N. Coll. Doat, vol. 150, and the original 1170s cartulary. The chronicle is published on page 371 (no. 470): this is the only lengthy authentic contemporary account of a Cistercian foundation for the region. Other surviving accounts of Cistercian foundations for southern France include the Bonneval chronicles—their authenticity is, however, questionable: see Rigal, *Cartulaire de l'abbaye de Bonneval-en-Rouergue*, ed. P.-A. Verlaguet, intro. J.-L. Rigal (Rodez: Carrère, 1938), xv ff., and 668–78; Bonnefont's foundation account in *Rec. de Bonnefont*, no. 2 (1136–38); an inscription for Aiguebelle in *Chartes et documents de l'abbaye de Notre Dame d'Aiguebelle. Commission d'histoire de l'ordre de Cîteaux II*, ed. anon. (Lyon: Audin, 1954), 1: no. 4 (1137); a text for Bonnevaux in *Cartulaire de l'abbaye Notre-Dame de Bonnevaux au diocèse de Vienne, ordre de Cîteaux*, ed. Ulysse Chevalier (Grenoble: F. Allier, 1889), no. 8 (1117); and a letter concerning the early hermits at Grandselve written to the king of England by a bishop of Toulouse in the early twelfth century which was copied on the fly-leaf of a manuscript in the Grandselve library: Toulouse, Bibliothèque municipale, MS 152, no. 6 (n.d.).

[9] Scattered documents found elsewhere have been published as *Cart. de Bonneval* (see n. 8 above); the fire in Bonneval's archives is described ibid., v–ix.

Valmagne (incorporated 1155). Rich archival materials document the early history of these two abbeys, both of which were incorporated rather than founded *de novo* by the order. Fontfroide, daughter-house of Grandselve, was an important link for the Clarevallian filiation line to Cistercian houses in Catalonia and reconquered Spain.[10] Valmagne, on the other hand, maintained ties with the newly powerful Guillelm family of Montpellier and with Provence, whose countess Cecilia (who had married a Trencavel viscount) was among the monastery's founders; Valmagne had fishing rights in the lagoons along the Mediterranean coast, extensive vineyards and pasture lands, and a varied group of granges.[11]

Farther east, near the mouth of the Rhône River, the Cistercian abbey of Franquevaux (founded in 1143) was located on the marshes of the Rhône delta or Camargue; it sited its granges in the vicinity of the cities of Nîmes and Arles, with which it had commercial ties, and sent its animals into the pasture lands of the Cavaillon east of the Rhône. Although a number of early documents on Franquevaux survive, its abbey church has virtually disappeared.[12] In contrast, there are virtually no surviving documents from the famous Provençal Cistercian houses of Sénanque (founded 1148), Silvacane (incorporated 1147), and le Thoronet (founded 1137), all located east of the mouth of the Rhône River, but all the churches still stand as monuments of early Cistercian architecture.[13] East of the Rhône and north of the county of Provence, in what was in the Middle Ages the marquisate of Provence, were Cistercian houses at Léoncel (founded 1137) and Bonnevaux (founded 1119) from which there are still considerable records. These last two houses practiced pastoralism in the high valleys of the upper Isère and Durance Rivers and in the Alps north of Grenoble, while concentrating their agriculture in the Rhône valley; Bonnevaux had strong ties to the north with such cities as Lyons.[14]

---

[10] On links between Fontfroide and Spain, see François Grèzes-Rueff, "L'abbaye de Fontfroide et son domaine foncier au xiie–xiiie siècles," *Annales du Midi* 89 (1977): 257–58. Few original documents survive, but two inventories of Fontfroide's archives contain nearly complete summaries of the early acts: Carcassonne, A. D. Aude, H211, "Inventaire de la manse conventuelle": and Narbonne, B. Mun., MS 259, "Inventaire de la manse abbatiale," which Mlle Denise Rouanet, municipal librarian of Narbonne, allowed me to photograph.

[11] See the Valmagne cartulary, available for consultation as microfilm no. 1 ms. 260–61, in Montpellier, A. D. Hérault, private deposit, especially section entitled, "Princes," vol. I, fols. 150 r ff., and "Cartulaire dit de Trencavel (XIe–XIIIe siècle)," which belongs to the Société Archéologique de Montpellier who allowed consultation of a microfilm of that volume also on deposit in the departmental archives of the Hérault.

[12] Nîmes, A. D. Gard, H33–H103, Franquevaux archives. Among these is one concerning the foundation of a hospice for Cistercians in Nîmes, ibid., H74 (1215), discussed in Constance H. Berman, "Monastic Hospices in Southern France," forthcoming. The village of Franquevaux still stands, but nothing is distinguishable of the church, which seems to have been converted into a wine cellar.

[13] Jean Barruol, *Sénanque et le pays du Lubéron au Ventoux* (Lyon: Lescuyer, 1975), nn. 17 ff.

[14] See the Bonnevaux reference in n. 8 above, and *Cartulaire de l'abbaye Notre-Dame de Bonnevaux au diocèse de Vienne, ordre de Cîteaux*, ed. anon. moine de Tamié (Notre-Dame de

Much of the prestige of the Cistercian order in the twelfth century was generated by Bernard of Clairvaux, whose journeys across Europe on ecclesiastical business in the 1140s attracted both individual recruits and whole monasteries to the order.[15] Bernard's insistent advocacy of the superior virtues of the new order and the reaction to his statements by the previously unchallenged Cluniacs developed into a rhetorical battle between the two orders, in which the polemics of both sides had little basis in reality and in which the similarity in goals of the two monastic groups was obscured by their argument over the proper means to achieve their common ends.[16] Debate over the legitimacy of their order made the Cistercians defensive and self-righteous in their description of their origins, and soon the themes of the debate became popular literary "topics," which were incorporated into the writings of friend and foe alike and often repeated by historians up to the present.[17] Frequently, Cistercians and Cluniacs were presented by narrators as opposite poles, as either end of a spectrum, or worse still, as the two sides of a balance on which whatever good said of one implied evil of the other. Moreover, the situation was confused by the fact that the stated ideals and purposes of the Cistercians, as presented by their twelfth-century advocates, could with only slight twists of language become denouncements of their enemies.

Since the economic practices of the early order are reported in similar terms by pro-Cistercian and anti-Cistercian writers alike, the very unanimity of the sources has tended to confirm their reliability. For instance, in their earliest collection of statutes, traditionally dated to 1134, the Cistercians

---

Tamié, Savoy, 1942), which contains a number of undated references to Bonnevaux's rights in Lyons, and *Cartulaire de l'abbaye Notre-Dame de Léoncel, diocèse de Die, ordre de Cîteaux*, ed. U. J. Chevalier (Montélimar: Bourron, 1869).

[15] On Bernard's role in incorporation throughout Europe, see Lekai, *Cistercians*, 33–51. An attempt by the order to incorporate the order of Chalais in Provence and Savoy was blocked by the papacy: see Jean-Charles Roman d'Amat, *Les Chartes de l'ordre de Chalais: 1101–1400* (Paris: Picard, 1923), 1: 12–14. A graph of Cistercian growth is given by R. W. Southern, *Western Society and the Church in the Middle Ages* (Harmondsworth, Eng.: Penguin, 1970), 254; that growth is mapped by R. A. Donkin, "The Growth and Distribution of the Cistercian Order in Medieval Europe," *Studia monastica* 9 (1967): 275–86; both draw their data from Leopoldus Janauschek, *Originum cisterciensium* (Vienna, 1877: repr. Englewood, N.J.: Gregg Press, 1964). Cistercian houses sometimes attributed their origin to Bernard's intervention even when Bernard could not possibly have been involved; for instance, Belleperche's attempt to associate a site change made after Bernard's death with his authority: Paris, B. N., Coll. Doat, vol. 91, Belleperche, fols. 7v–11r (1166). As described by the Doat copyist this is an "extrait d'une copie en parchemin trouvée aux archives de l'abbaye de Belleperche."

[16] On the Cistercian/Cluniac debate, see Knowles, *Monastic Order*, 224–26; idem, *Cistercians and Cluniacs* (New York: Oxford University Press, 1955), passim; and Idung de Prufening, *Cistercians and Cluniacs: the Case for Cîteaux*, ed. J. O'Sullivan, et al. (Kalamazoo, Mich.: Cistercian Publications, 1977); on tithes, see Giles Constable, "Cluniac Tithes and the Controversy between Gigny and Le Miroir," *Revue bénédictine* 70 (1960): 591–624.

[17] Anti-Cistercian polemic is discussed by Coburn V. Graves, "The Economic Activities of the Cistercians in Medieval England. (1128–1307)," *Analecta sacri ordinis cisterciensis* 13 (1957): 45–54; complaints about the order's tithe ownership in Giles Constable, *Monastic Tithes from their Origins to the Twelfth Century* (Cambridge: Cambridge University Press, 1964), esp. 270 ff.

corroborate other narrators' descriptions of the order which founded its houses "in horrible places" and "far from human habitation."[18] The economic practices of the early order are described by early Cistercians themselves in the following items from that collection:

One: Where houses must be constructed: None of our houses must be constructed in cities, castles, or villas, but in places far from human society.

Two: The source of food for the monks ought to be: Monks of our order ought to produce food by manual labor, by cultivating land, from stock-raising, for which we are allowed to possess for our own use: streams, woodlands, vineyards, meadows, lands remote from the dwellings of secular men, and animals. . . . For using, raising, and maintaining these (whether nearby or far-off, but not farther than a day's journey), we can have granges managed by lay-brothers.

Eight: Concerning the lay-brothers: The work at the granges must be done by lay-brothers and hired laborers, . . .

Nine: What [sources of] income we do not have: The disposition of our order prohibits [ownership of] churches, altars, burials, tithes from the agricultural work of others and/or from [their] stock-raising, *villae*, serfs, rents from land, income from ovens and mills, and other income similarly counter to monastic purity.

Twelve: How a new church is provided with abbot and monks and other necessities: Twelve monks with a thirteenth as abbot are sent out to the new "coenobia": nor are they sent there until the place is provided with books, houses, and other necessities. . . .[19]

Other Cistercian sources describe in similar terms the "deserted places" in which the order's new houses were founded, the poverty of the early monks, the miracles which provided them with food, or how they scratched out an existence in brush-covered, forested lands in the company of wild beasts, . . .[20]

---

[18] According to the Silvanès chronicle, "Locum habitabilem ex inhabitabili reddiderunt," *Cart. de Silvanès*, p. 382. The allusions are from the Vulgate, "in locus horroris et vastae solitudinis," Deut. 32:10, and "de deserto venit, de terra horribili," Isaiah 21:1.

[19] Author's translation. I. *Quo in loco sint construenda coenobia:* In civitatibus, castellis, villis, nulla nostra construenda sunt coenobia, sed in locis a conversatione hominum semotis [sic].

V. *Unde monachis debeat provenire victus:* Monachis nostri ordinis debet provenire victus de labore manuum, de cultu terrarum, de nutrimento pecorum, unde et licet nobis possidere ad proprios usus aquas, silvas, vineas, prata, terras a saecularium hominum habitatione semotas [sic], et animalia. . . .Ad haec exercenda, nutrienda, conservanda, seu prope seu longe, non tamen ultra dietam, grangias possumus habere, per conversos custodiendas.

VIII. *De conversis:* Per conversos agenda sunt exercitia apud grangias et per mercenarios, . . .

IX. *Quod redditus non habeamus:* Ecclesias, altaria, sepulturas, decimas alieni laboris vel nutrimenti, villas, villanos, terrarum census, furnorum et molendinorum redditus, et cetera his similia monasticae puritati adversantia, nostri et nominis et ordinis excludit institutio.

XII. *Quomodo novella ecclesia abbate et monachis et ceteris necessariis ordinetur:* Duodecim monachi cum abbate terciodecimo ad coenobia nova transmittantur: nec tamen illuc destinentur donec locus libris, domibus et necessariis aptetur, . . . *Statuta capitulorum generalium ordinis cisterciensis ab anno 1116 ad annum 1786*, 8 vols., ed. J.-M. Canivez (Louvain: Bureaux de la Revue, 1933), vol. 1 (1134), *prima collectio*, pp. 12–16.

[20] See the Silvanès chronicle, cited above in n. 8, or the account of the foundation of Obazine in western France: *Vie de Saint Etienne d'Obazine*, ed. Michel Aubrun (Clermont-Ferrand: Institut d'Etudes, 1970), book I, part 4.

Contemporary critics of the Cistercians tended to use the same images. Often cited is Gerald of Wales, whose "fair appraisal of the two orders" (Cistercian and Cluniac) has been considered particularly reliable because he is critical of both groups.[21] He describes the Cistercians as land-hungry, "constantly on the look-out for rich lands and broad pastures," but the Cluniacs were wastrels in comparison:

> Give the Cluniacs today a tract of land covered with marvelous buildings, endow them with ample revenues and enrich the place with vast possessions; before you can turn around it will all be ruined and reduced to poverty. On the other hand, settle the Cistercians in some barren retreat which is hidden away in an overgrown forest; a year or two later you will find splendid churches there and monastic buildings, with a great amount of property and all the wealth you can imagine.[22]

Not surprisingly, this picture of how Cistercian monasteries transformed wastelands into new Edens as presented by medieval advocates and critics alike has generally been accepted by historians and incorporated into the standard description of the order's early years.[23] To summarize the main events of the textbook account: at the urging of bishops and lords, and after appeals from the community, a group of monks would set out from Burgundy in the proper apostolic number (six or twelve plus an abbot) to make a new foundation. After a time of wandering they would discover a forested and secluded site (often owned by those who had appealed to them to make the foundation). There they would settle, building huts in the forest, their inadequate food supplies supplemented by miraculous intervention. They would cut down trees, begin their stock-raising, and by their own hands erect the simple churches which they dedicated to the Virgin.[24]

Until recently, exceptions to this ideal description were often discounted as regional aberrations or those introduced by incorporated groups; some articles have even apologized that the archival evidence does not document the picture of early Cistercian foundations which scholars have anticipated.[25] With this idealized picture, moreover, economic and social histo-

---

[21] Gerald of Wales, *The Journey through Wales*, trans. Lewis Thorpe (Harmondsworth: Penguin, 1978), Book I, chap. 3, p. 103.

[22] Ibid., 106.

[23] The modern textbook description of the Cistercian foundation and economy is found, for example, in the work of such well-known scholars as M. M. Postan, *The Medieval Economy and Society* (Harmondsworth: Penguin, 1975), 102, and N. J. G. Pounds, *An Economic History of Medieval Europe* (London: Longman, 1974), 171–73, as well as in more general studies as those by Hugh Trevor-Roper, *The Rise of Christian Europe* (London: Thames and Hudson, 1963), pp. 121–22; R. S. Hoyt, *Europe in the Middle Ages*, 2nd ed. (New York: Harcourt, Brace and World, 1966), 265, and C. W. Hollister, *Medieval Europe: A Short History* (New York: Oxford University Press, 1968), 169.

[24] *Cart. de Silvanès*, 371 ff., esp. 382: "In quo casulas propriis manibus fabricantes mansuerunt, bestiis sociati, quotidiano tamen labori insistentes, dumeta falcibus resecantes, terram ligonibus proscindentes, locum habitabilem ex inhabitabili reddiderunt."

[25] Inference from "present-day" appearance of Cistercian sites or from the order's legislative documents can be found in H. de Barrau, "Etude historique sur l'ancienne abbaye de Bonnecombe," *Mémoires de la société des lettres, sciences et arts de l'Aveyron* 2 (1840): 193, or abbé

rians have developed theories about the contribution of the early Cistercians to medieval economic growth. These models have been constructed without questioning whether the descriptions of Cistercians employed are reliable, given that such descriptions are drawn from propagandizing and polemical accounts. Often such models have been influenced by theories regarding the history of the American West; medieval economic historians in Europe and America alike developed a "frontier thesis" of early Cistercian expansion which described how Cistercian monks, like American pioneers, went out into the wilderness—to uninhabited, never-before cultivated land—and converted forest to arable land which was suitable for cultivation. These historians claimed that the Cistercians, like the prairie settlers of the new world, derived unexpectedly large yields from rich soils which had never before been cultivated. The yields from these "bumper crops," plus savings derived from Cistercian frugality and manual labor in the fields, explained the rapid expansion and growth of the Cistercian order in the twelfth century. According to this "frontier" model the Cistercians, by seeking a life of asceticism and apostolic poverty, were twelfth-century predecessors of Weber's Puritans, and Cistercian monks, like those later Protestants, were for their efforts ironically heaped with riches.[26]

Bousquet, "Anciennes abbayes de l'ordre de Cîteaux dans la Rouergue," *Mémoires de la société des lettres, sciences, et arts de l'Aveyron* 9 (1860–67): 1–2. Similarly, R. Rumeau, "Notes sur l'abbaye de Grandselve," *Bulletin de la société de géographie de Toulouse* 19 (1900): 248; B. Alart, "Les monastères de l'ancien diocèse d'Elne," *Mémoires de la société agricole des Pyrénées-orientales* 11 (1872): 280; and more recently, J. Laurent, "Les noms des monastères cisterciens dans la toponymie européenne," *Saint Bernard et son temps* (Dijon: Association, 1928, 1: 171). On the regret that surviving documents do not illustrate an abbey's anticipated early origins, see Elizabeth Traissac, "Les abbayes cisterciennes de Fontguilhem et du Rivet et leur rôle dans le défrichement médiéval en Bazadais," *Revue historique de Bordeaux* 9 (1960): 141–58, a study which adduces little or no evidence of Cistercian clearance. On a "mitigated" Cistercian rule, see L. de Lacger, "Ardorel," *Dictionnaire d'histoire et de géographie ecclésiastique* (1924), 7: cols. 1617–20. On the contention that decadence was introduced by incorporated houses, see Bennett D. Hill, *English Cistercian Monasteries and their Patrons in the Twelfth Century* (Urbana, Ill.: University of Illinois Press, 1968), esp. chap. 4, "The Congregation of Savigny and its Impact on the Cistercian Order."

[26] James W. Thompson, *The Economic and Social History of the Middle Ages* (New York, [1928]; repr. Ungar, 1959), 611, says, "The Cistercians deliberately sought out remote and isolated regions in Europe and these naturally were in forests and marshy locations." Henri Pirenne in vol. 1 of his *Histoire de Belgique*, 5 vols., 5th ed. (Brussels: Lamertin, 1929), 301, described the Cistercians as "established almost always in uncultivated lands, in the middle of woods, heath and marshes, they vigorously gave themselves over to clearance." Marc Bloch, *French Rural History: An Essay on its Basic Characteristics*, trans. from 1931 French edition by Janet Sondheimer (Berkeley: University of California Press, 1970), 14–15, described the Cistercian monastic property: "the abbey itself was always situated in some place remote from human habitation—usually a wooded valley . . . and the subsidiary 'granges' also avoided the neighborhood of peasants' dwellings. They were established in 'desert' solitudes." Bloch's description echoes the Cistercian *prima collectio* (quoted in n. 19 above), but see also Georges Duby, *The Early Growth of the European Community*, trans. Howard B. Clarke (Ithaca: Cornell University Press, 1974), 219–20, who gives a description of the order similar to Bloch's; elsewhere Duby takes a slightly different view, see n. 28 below. American proponents of the frontier thesis, following Thompson in general, include Bryce Lyon, "Medieval Real Estate Development and Freedom," *American Historical Review* 63 (1957): 47–61, and Archibald R. Lewis, "The Closing of the Medieval Frontier, 1250–1350," *Speculum* 33 (1958): 475–83.

It is now becoming accepted that the early Cistercians, anxious to justify their privileges and to prevent criticism of the irregularities of the foundation at Cîteaux, stridently insisted in their narrative and legislative documents on the poverty, isolation, and self-sufficiency of their houses, and on the foundation of their abbeys in areas of otherwise unwanted lands.[27] As the order's early accounts become recognized as propaganda, more historians will reject their use for descriptions of the order's economic practice. Recent studies, by close examination of the land acquisition documents, have confirmed that a reevaluation is necessary, because traditional notions about the order and its role in economic development are distorted.[28]

Earlier ideas about Cistercian foundations in wilderness areas, however, and the economic advantages to be reaped from the order's assumed "frontier" activities, cannot simply be rejected. They must be replaced by new and equally adequate explanations of the order's quickly gained wealth and prestige. Otherwise, the previous economic interpretations will remain popular because they seem to work well in explaining how Cistercian houses became wealthy and how the Cistercians could grow from a single, semi-eremitical abbey in Burgundy to a powerful order of over five hundred houses in the course of a century.[29] Older models will only be dismissed

---

[27] Efforts in the 1950s to challenge the authorship of certain early Cistercian narratives were made and met with much the same vitriolic language and polemic which had characterized the twelfth-century debate; see David Knowles, "The Primitive Cistercian Documents," *Great Historical Enterprises: Problems in Monastic History* (London: Thomas Nelson and Sons, 1963), 197–222. The controversy had not yet been resolved in the late 1970s when Lekai wrote *Cistercians.* See p. 22, where he says, "it is still impossible to replace the old, traditional image of early Cîteaux with a similarly neat and clear picture sketched with the assistance of modern scholarship." Studies of ideals and reality have included: Richard Roehl, "Plan and Reality in a Medieval Monastic Economy: The Cistercians," *Studies in Medieval and Renaissance History* 9 (1972): 83–113, and Louis J. Lekai, "Ideals and Reality in Early Cistercian Life and Legislation," *Cistercian Ideals and Reality,* ed. John R. Sommerfeldt (Kalamazoo, Mich.: Cistercian Publications, 1978), 4–19, and see p. 33. n. 20 below.

[28] Georges Duby, *Rural Economy and Country Life in the Medieval West,* trans. Cynthia Postan (Columbia, S.C.: University of South Carolina Press, 1968), 70–71: "Monks were not the chief architects of the movement of reclamation as has for so long been believed. . . . Wherever detailed research has been carried out it has shown that these new monasteries [Grandmont and Cîteaux in particular here] were set up in clearings already prepared to a greater or lesser extent. . . . [I]n actual fact, the only holy men who effectively helped with their own hands to wage the assault against the wastes, who felled trees and ploughed up new arable, were the hermits who at that time lived in great numbers on the fringes of the European forest." Duby's remarks, published in France in 1962, were paralleled in the 1963 publication by B. H. Slicher van Bath, *The Agrarian History of Western Europe, 500–1850,* trans. Olive Ordish (London: E. Arnold, 1963), 153–54, who says: "Whether the young monastic order of Cistercians had an important share in the land reclamation of western Europe is an open question. . . . Recent research . . . in lower Saxony has revealed that the monasteries there were endowed with land already under cultivation and that the monks had no part in reclaiming it." This is certainly the impression given by more recent studies such as those cited in n. 29 below, and it is from studies such as these that Fossier and Higounet draw their conclusions; see Preface, note 1.

[29] The old textbook account does not fit the archival evidence for the individual houses of the order either in southern France or in several other parts of Europe. Indeed, even in the Cistercian houses in Burgundy—the region for which the traditional model might be considered most applicable—there were early divergences from the ideal. See R. A. Donkin, *The Cistercians:*

when replaced by ones which better account for this early growth. Here an attempt is made to advance a new economic explanation for the Cistercian order's early and phenomenal expansion in southern France. It is based not on "bumper crops," or assumed high fertility in "new lands," or a "frontier," but on the coincidence of managerial prowess, interdependent agriculture and pastoralism, direct cultivation of newly created granges, land acquisition tied to the recruitment of *conversi* or lay brothers as laborers, the availability of markets in neighboring cities for surplus products (with market privileges which gave the order an edge over competitors), and exemption from ecclesiastical tithes. These advantages for Cistercian agricultural production and marketing in twelfth- and thirteenth-century southern France explain Cistercian economic expansion and success. The managerial efforts of the Cistercians allowed a more rationalized, more up-to-date agriculture on their vast twelfth-century corporate farms in southern France. Such rationalization of agriculture was probably as important to the region's economic growth as the clearance and reclamation with which the order is often credited would have been. Cistercians in southern France were not the "frontiersmen" or "pioneers" which the "frontier thesis" of early Cistercian settlement would lead us to believe, but instead were entrepreneurs and managers. By adapting the new ideas about agriculture in the order's ideology to local conditions, Cistercians did introduce a new type of monastic agriculture: direct cultivation on large, compact granges, with *conversi* laborers and a variety of tax exemptions. These granges were by and large, however, created out of previously settled lands. In southern France such adaptation to local settlement conditions was crucial, for it allowed the introduction of a reformed monastic presence, which was badly needed in that region where heresy was rife, but where there was little "new" land or forested wilderness with which to endow the "ideal" Cistercian house.[30]

Studies in the Geography of Mediaeval England and Wales (Toronto: Pontifical Institute of Mediaeval Studies, 1978); Brian Patrick McGuire, *The Cistercians in Denmark: Their Attitudes, Roles and Functions in Medieval Society* (Kalamazoo, Mich.: Cistercian Publications, 1982), esp. 37–111; Werner Rosener, "Bauernlegen durch Klösterliche Grundherren im Hochmittelalter," *Zeitschrift für Agrargeschichte und Agrarsoziologie* 27 (1979): 60–93; and Walter Schlesinger, "Flemmingen und Kühren zur Siedlungform niederländischer Siedlungen des 12. Jahrhunderts im Mitteldeutschen Osten," *Die Deutsche Ostsiedlung des Mittelalters als Problem der Europäische Geschichte*, Reichenau: Vorträge, 1970–1972, ed. Walter Schlesinger (Sigmaringen: Thorbecke, 1975), 263–309. For Burgundy, see *Le Premier Cartulaire de l'abbaye cistercienne de Pontigny (XIIe–XIIIe siècles)*, ed. Martine Garrigues (Paris: Bibliothèque nationale, 1981), 19ff.; *Chartes et documents concernant l'abbaye de Cîteaux: 1098–1182*, ed. J.-M. Marilier (Rome: Editions cistercienses, 1961), esp. no. 41, part 5; no. 48; no. 51, part 5; no. 56, etc.; *Recueil des pancartes de l'abbaye de la Ferté-sur-Grosne, 1137–1178*, ed. Georges Duby (Aix: Publ. de la Faculté de lettres, 1953), 18–21; and Robert Fossier, "L'essor économique de Clairvaux," *Bernard de Clairvaux*, préface de Thomas Merton (Paris: Editions Alsatia, 1953), 95–114.

[30] On heresy in southern France, see Walter L. Wakefield, *Heresy, Crusade and Inquisition in Southern France: 1100–1230* (Berkeley: University of California Press, 1974); and John H. Mundy, *Liberty and Political Power in Toulouse: 1050–1230* (New York: Columbia University Press, 1954).

II. THE QUESTION OF NEW LANDS AND RECLAMATION BY CISTERCIANS

Recent studies of agricultural settlement in western Europe, such as that by Pierre Toubert on central Italy, suggest that the distinction between cultivated and uncultivated land in medieval agriculture was more vague than is generally admitted—that it is incorrect to think of land as having been either totally inhabited or totally uninhabited, since the entire spectrum of conditions of occupation existed in the medieval rural world.[1] Probably the regions into which the Cistercians came should not be viewed as having had pockets of dense settlement surrounded by waste, at least for Mediterranean regions by the twelfth century. In any case, it is clear from the charters of Cistercian monasteries in southern France that properties acquired by the order in that region were located almost entirely in already settled, cleared, and cultivated areas, not in wilderness. If twelfth-century religious men in southern France undertook the kind of reclamation which has generally been attributed to the Cistercians, those "pioneers" were hermits, not Cistercians. Indeed, in many places in that region Cistercian abbeys had been preceded by hermitages, or monasteries founded by earlier reformers.[2]

The archival documents give no suggestion that Cistercians in southern France created either abbey- or grange-sites by clearance or drainage, for even vague suggestions of such activities are almost nonexistent in surviving documents. Although it is impossible to contend that Cistercians did absolutely no clearance, in general abbeys seem to have been more anxious to preserve forests than to cut them down.[3] Charters from throughout the region confirm that the land which the order came to own in southern France was most often in territory which had been occupied and cultivated for a number of previous generations, and that Cistercians in the Midi acquired their land almost entirely without undertaking any pioneering activities. This conclusion is based on a wealth of evidence showing that land coming into Cistercian hands was already under cultivation at the time of its acquisition by the monks. This conclusion stands on its own, but it is also reinforced by the argument from silence—that the documents

---

[1] Pierre Toubert, *Les structures du Latium médiévale; Le Latium méridional et la Sabine du IXe siècle à la fin du XIIe siècle*, 2 vols. (Rome: Ecole Française, 1973), 1: 181 ff.

[2] See quote from Duby, *Rural Economy*, above in Chapter One, note 28; Table 1 below, and Berman, "Growth," forthcoming.

[3] The need to preserve forest is underlined by Fossier, "L'économie cistercienne du nord-ouest," 69.

almost never record grants of permission to cut down trees and brush or drain marshes, and rarely even suggest the conveyance of absolute ownership of (as opposed to usage of rights in) forested lands. Such documents also show that the land acquired by the Cistercians in southern France was generally not marginal land, or was marginal only in the sense that increasing fragmentation was making it less and less profitable.

The previously cultivated condition of land coming into Cistercian hands is revealed by surviving documents in a number of ways: first, in the traditional terminology used to describe land units conveyed to the order; second, in references to established ecclesiastical tithes on that land; third, by references to peasant occupants of that land; fourth, by the obvious fragmentation of the land units conveyed, which suggests a long history of settlement; fifth, by the records of large sums paid by the monks to the land's previous owners and cultivators; and finally, by references to "development"—to the existence of planted vines, houses, and mills on properties acquired by the order. Such indices are found in an overwhelming majority of southern-French charters. They show not only that reclamation there was not undertaken by the Cistercians, but also suggest that where such activities were undertaken by religious groups, the heavy labor had usually been assigned to dependent-tenant laborers.[4] As will be seen in chapter 3, the Cistercians in southern France, in conformity with the ideals of the early order and as part of the land consolidation process, sought to rid their land of those dependent tenants who were used by other groups in land clearance and reclamation. Moreover, in southern France, reclamation had already been completed in most areas by the predecessors of the Cistercians, and the order added little "new" land to their acreages by reclamation. The monks of the Cistercian order acquired their extensive properties by purchase or gift from earlier owners and cultivators, not by reclaiming land. They were too late on the scene to have many opportunities to undertake clearance of land, and in general, the lack of dependent-tenant farmers left them ill-provided with laborers to do so.

Although the terminology tends to be repetitive and almost formulaic, the descriptions of land conveyed to the order found in Cistercian charters provide particularly clear evidence that land coming into the order's ownership had been occupied and cultivated prior to its acquisition by the monks. Although the terms in individual acts changed with circumstances and from locale to locale, several examples of charter phraseology can demonstrate its explicit nature. For the Rouergat abbey of Bonnecombe, for instance, charters of land conveyance describe the rights transferred

---

[4] Clearance documented for Cistercian monks was slow, as Fossier points out, loc. cit., and was also so late that by that time tenant-farmers may have been being used. There is evidence of reclamation by Benedictine houses in the eleventh century, and by thirteen-century communities of nuns who had peasants to work for them. See Higounet, "L'occupation du sol du pays entre Tarn et Garonne au moyen âge," *Annales du Midi* 68 (1956): 227–254 on eleventh-century Benedictines, and Berman, "Examples from the Diocese of Chartres: Economic Activities of Medieval Cistercian Houses for Women," forthcoming.

by using the traditional terms for established tenures. For example, one donor conveyed the entire *allod*, vicarial rights, and whatever else he has in the two *mansi* of Vareilles with all the appurtenances of those *mansi*, and similar rights in the *mansus* of la Coste and in the *caputmansus* of la Coste with all their appurtenances; each *mansus* owed *hospitium* for five knights as well as a pig, a sheep, twelve *nummi*, and ten *sestiers* of rye annually.[5] Similarly an act of 1180 conveyed to those same monks of Bonnecombe all that Ato of Cums held in seven *mansi* at Riparia, la Coste, Vareilles inferior and superior, Puy, Martres, and Berengairenc. This included the quarter part of existing rents on meat, wool, and cheese in those seven *mansi*, rights to the *fevum* and other *censuales*, and the *quartum* of wheat in the *mansus* of Puy, a sixth of the wheat in the *mansus* of Berengairenc, the tithe of beasts in the *mansi* of Vareilles when the tithe was paid there, and all rights in the *appendaria* of Becerières which Ademar Raoul held. For this gift, the donor, Ato of Cums, received one hundred *solidi* and was granted confraternity in the benefits of the monastery with his parents.[6]

Similarly in Gascony, Jourdain of Marrast gave the monks of Berdoues "all the cultivated and uncultivated land which he had in the territory of Berdoues and in its appurtenances on either side of the Baise River, in addition to all that which he held in the grange between Berdoues and St. Mavius, with ingress and egress, water, pasture and woods"; for this he received thirty *solidi* of Morlaas and the promise that he would be received as a monk in the monastery if he came within five years.[7] An example

---

[5] *Cart. de Bonnecombe*, no. 251 A (1163): ". . . Ego Arnaldus de Taurinis et ego Poncia, soror ejus, per nos et per omnes successores nostros, in perpetuum, bona fide, sine dolo et sine omni retentione, donamus et concedimus pro salute nostra Deo at Beate Marie Candelii et Gauzberto, abbati, et vobis fratribus ejusdem loci presentibus atque futuris totum alodium et vicariam, et quicquid aliud habemus et habere debemus vel alius per nos in mansis de Vallelas cum pertinentiis suis, et in manso de Costa, et in capmas cum omnibus pertinentiis suis, ut libere et quiete habeatis et possideatis perpetuo jure; et sciendum quod propter hoc habuimus a vobis C solidos caritati[ve]. Testes sunt: . . . —Memorie commendandum quod unumquodque mansorum supradictorum debebat hospitium V militum et porcum et multonem, unumquodque XII nummorum et X sextarios de siligine.—. . . M.C.LX.III."

[6] Ibid., no. 260C (1163): ". . . M.C.L.XXX. Ego Ato de Cums, . . . dono Deo et monasterio Sancte Marie Bonecumbe et Poncio, abbati ejusdem loci . . . eterna donatione omne quod habeo et habere debeo vel aliqua persona per me in his VII mansibus, scilicet in Riparia, in Costa, in Vallellas inferiores, et in Vallellas superiores, in Pogeto, in Martors, in Berengairenco, quod est quarta pars de carnenco in omnibus his VII mansibus et de lana et de caseis, et feuum et guarbe censulaes qui ibi actenus habui, et quarta pars blati in Pogeto, et sexta in Berengairenco, et decimum bestiaris eorum in manso Vallis, cum ibi paverit. Dono . . . omne jus quod habeo in apendaria, quam tenet Ademarus Radulfus in Beceria; et vos donastis michi in caritate pro his centum solidos, quorum nichil vobiscum remansit in debito."

[7] *Cart de Berdoues*, no. 7 (1161): "Sciendum est quod Jordanus dictus filius Gualardi, filius Jordani et Gassiarnaldi de Marrast, bono animo et bona voluntate, bona fide et sine omni retentione pro se et pro omnibus successoribus suis presentibus et futuris donavit, concessit et absolvit Deo et beate Marie Berdonarum et Arnaldo abbati et conventui ejusdem loci presenti et futuro totam terram suam cultam et incultam quam habebat vel habere debebat per se vel per aliam personam in territorio Berdonarum et in omnibus pertinenciis ejus ante Baisiam et retro Baisiam et hoc quod habebat ad portam grangie que est inter abbatiam Berdonarum et sanctum Mavium. Totum hoc quod infra has predictas adjacentias predictus Jordanus habebat, donavit predicto abbati et predicto conventui pro amore Dei et redemptione anime sue et

from a somewhat later act for Berdoues includes the description of landholdings standard in the charters of the south. In this conveyance, dated 1215, Arnold of Oms gave his own person and his son to the house of Berdoues to become *confratri* (lay brothers?) there; he also gave his rights in the castle of Serres and the territory of Oms, described as: "lordship and rents, men and women there, land in holdings which were cultivated or not cultivated (*in terris cultis et incultis*), pasture, water rights, hunting rights, and all other rights pertaining to those properties."[8]

Of particular note here is the recurring formula: *in terris cultis et incultis* which appears not only in this charter of 1215, but in many earlier charters from the Midi. In the absence of any contemporary gloss explaining the use of such a term, the word *incultus* must be considered in context.[9] In the documents studied here, it signifies one of the normal appurtenances of settled agricultural land, probably either land lying fallow, or an area of "out-field" or wasteland used as pasture and as a source of forest products and construction materials. What the adjective *incultus*, in company with the term *cultus*, does *not* mean in such charters is some kind of previously unemployed territory or "new" land awaiting settlement. Instead, it implies parts of the continuum mentioned earlier—between kitchen garden and forest—which would gradually have extended through less and less intensively used, but nonetheless essential, holdings: cultivated fields, orchards, pastures, then "waste" and finally "forest" which was managed, selectively cut, and protected, perhaps even replanted. All these types of land were essential to the village economy, and to the monastic economy as well. Only when *incultus* is found outside the formulaic phase *in terris cultis et incultis*, might the term possibly refer to totally unexploited lands. Such a usage is not found in southern French Cistercian charters.[10] Con-

---

parentum suorum, cum ingressibus et egressibus, aquis, pascuis et nemoribus, ut habeant et possideant libere et quiete sine omni sua et suorum contradictione in perpetuum; et debet inde facere bonam et firmam garentiam de omnibus amparatoribus predictis habitatoribus Berdonarum et pro hoc predicto dono predictus abbas donavit predicto Jordano ex caritate XXX sol. morlanorum." It continues: "Et convenit ei quod si infra V annos ad religionem venerit, in Berdonis pro monacho vel converso recipiatur. Si autem transacto spatio V annorum reliquerit seculum debet recipi pro converso tantum non pro monacho." Such recruitment from among the local populace is discussed in more detail in chapter 3.

[8] "Manifestum sit omnibus hominibus qui hec audierint quod Arnaldus de Oms dedit seipsum et filium suum per confratres Deo et beate Marie Berdonarum et Guillelmo abbati et conventui ejusdem loci presenti et futuro. Dedit et concessit et tradidit totum hoc quod habebat et habere debebat per se vel per aliam personam aliquo modo in castro de Serris et in omnibus terminiis ejus et ad Oms et in omnibus pertinentiis suis, scilicet in dominiis, in usaticis, in hominibus et feminis, in terris cultis et incultis, in pascuis, in aquis et venationibus et in omnibus aliis rebus ad ipsum pertinentibus." *Cart. de Berdoues*, no. 151 (1215).

[9] The standard medieval Latin dictionary defines "incultibilus" as "incapable of being cultivated," but does not include "incultus"; see C. du Fresne DuCange, *Glossarium mediae et infimae Latinitatis*, ed. Léopold Favre (Paris: Librairie des sciences et des artes, 1937–38), 4: 335. A more recent dictionary, J.-F. Niermeyer, *Mediae Latinitatis Lexicon Minus* (Leiden: E. J. Brill, 1974), gives no definition for "incultus," although "cultus" is defined as "reclaimed land": 287. This definition is too restricted for the term's usage, at least in southern France.

[10] Even in charters from Burgundy concerning Cîteaux, the use of "incultus" with this meaning is rare, but see for example in the early documents of *Chartes et documents concernant l'abbaye de Cîteaux, 1098–1182*, ed. J.-M. Marilier, no. 41 (early twelfth century):

veyances from that region mentioning *cultus et incultus*, like the example cited above, generally also mention men and women who were currently occupying the land conveyed, and the existence of nearby castles, tithes, or "improvements" to the land. All such accoutrements of the land as described in the charters contradict any conclusion that this was "new" land.

There are other traditional terms found in Cistercian charters similarly signaling the conveyance of established peasant farms to the order. For example, the terms describing a typical peasant holding or farm were *mansi, casalia,* or *tenementa,* depending on the region, but use of any of these terms shows that the order was acquiring previously cultivated tenures, occupied at the time of purchase by settled peasants. This is so, for instance, in the Rouergue and the Albigeois, where farms conveyed to the Cistercians were called *mansi, caputmansi,* and *appendariae* (the last term indicating already old additions to *mansi* and *caputmansi*).[11] A count of the frequency of such terms in the Rouergat Cistercian cartularies indicates that the bulk of land coming to the monks in that province was already settled. Among Rouergat charters for the abbeys of Bonneval, Bonnecombe, and Silvanès (altogether about 1500 acts for the years up to 1250), forms of the terms *mansus, caputmansus,* and *appendaria* appear in 43 percent of all acts concerning Bonneval, 63 percent of all acts concerning Bonnecombe, and 52 percent of all acts concerning Silvanès. These indices of previously occupied land coming into Cistercian ownership increase to 56 percent for Bonneval, 79 percent for Bonnecombe, and 69 percent for Silvanès, if charters containing indices of "development" or "improvement" of that land, such as references to existing vineyards, buildings, and castles, or to established tithes, are also counted.[12] Remaining acts for those three abbeys (the 44 percent, 21 percent, and 31 percent respectively which do not include such terms) do not thereby indicate clearance and reclamation; indeed, almost none of them could be thus interpreted. The remaining charters are simply less explicit: often confirmations of previous conveyances, or conveyances of "all land" in a particular area, or donations of pasture rights.

"I: Notum cupimus fore et presentibus et subsequentibus sancte Matris nostre Ecclesie filiis, Aimonem et conjugem ejus, Waronem quoque et Widonem filios ejusdem conjugis ac Dodonem et uxorem ejus natosque, illorum, monachis sancte Marie Novi Monasterii octo jornales terre inculte apud Gilliacum fundum concessisse. . . ."

"III: Monachi vero sancti Germani de Parisiaco quatuor jornales terre similiter inculte, insuper etiam nemorose, que contigua extat supra memoratis octo jornalibus habentes, antedictis monachis pauperibus Novi Monasterii causa Dei magneque sue misericordie communi consilio capituli sui contulerunt. . . ."

[11] See Bonnecombe cartularies, passim, as cited in next note. On the *appendaria,* see DuCange, *Glossarium* 6: 94, where it is described as an "appendix." An example of its use in the Rouergue is the *appendaria* of Parlant, which had tithes in a different category (*novales?*) not initially conveyed to Bonnecombe when other tithes in the *villa* of Magrin were received: Rodez, Soc. des lettres, Bonnecombe, "Cart. de Magrin," no. 57 (1182). Pasture rights at Parlant were similarly excluded in donations of other rights: *Cart. de Bonnecombe,* no. 165 (1174).

[12] *Cart. de Silvanès,* nos. 1–506; *Cart. de Bonneval,* nos. 1–162 and appendices; *Cart. de Bonnecombe,* passim, and Rodez, Soc. des lettres, Bonnecombe, "Cart. de Magrin," "Cart. de Moncan," "Cart. d'Iz et Bougaunes."

In Gascony, the traditional term for such settled peasant holdings was *casalium* or *casal*. Again, gross numerical trends gleaned from local Cistercian documents show how frequently land conveyed to the order was located in long-settled areas. For instance, the 825 acts in the Berdoues cartulary include references to over 230 different *casaux* coming into the hands of the monks; many of those *casaux* are mentioned in more than one charter.[13] For Gimont, Berdoues's daughter-house, similar numerical evidence can be cited which suggests that land was already occupied when it came into Cistercian ownership, since conveyances of *casaux* are frequently mentioned in its charters. For example, of the 214 charters for Gimont's grange of Laurs, 52 explicitly mention *casaux*, 18 mention rents, fields, lordship and tenurial rights over land, 10 concern pasture rights, and 44 mention that holdings were in the parish or territory of a particular church. Less explicit acts in that section of the Gimont cartulary indicate holdings in "territories" which were probably *villae* or parishes. Even without considering such "territories," nearly half of all documents for this grange explicitly refer to acquisitions in areas of established settlement.[14]

Similar "quantitative" proof of previous cultivation and settlement could be cited for any of the other abbeys of the Cistercian order in southern France. As in the examples from the Rouergue, when conveyances in Gascony do not mention established tenures or improvements on land it is because such donations and sales to the Cistercians were expressed in the vague terms quite typical of the contracts for this region west of Toulouse near Auch. A number of charters from Berdoues and elsewhere simply mention the donation or sale of "all my lands" or "all my rights" at a particular place without further clarification.[15]

Acquisition of existing ecclesiastical tithes is a second index of the long history of settlement on lands acquired by the Cistercians in southern France. There are many references in local documents to acquisitions in the *decimariae* or tithe-producing areas of existing parish churches, as well as conveyances to the Cistercians of those tithes and churches themselves. For example, in Provence the monks of Bonnevaux acquired tithes on previously cultivated lands in the parishes of St. Maurice and Landrins among many others.[16] In Languedoc, the abbey of Valmagne acquired rights to tithes and other properties in numerous parishes: among them, those of

---

[13] See *Cart. de Berdoues*, indices.

[14] *Cart. de Gimont*, II, passim.

[15] For example, the mortgage in *Cart. de Berdoues*, no. 134 (n.d.): "Sciendum est quod Bernardus de Mesplede . . . misit in pignus Alberto abbati Berdonarum . . . totam ex integro terram cultam et incultam quam habebat vel habere debebat per se vel per aliam personam a Lavazencs pro C sol. morlanorum . . ." or the conveyance ibid., no. 38 (n.d.): "Sciendum est quod Beliarda de Comas . . . donavit, concessit et absolvit Deo et beate Marie Berdonarum et Alberto abbati . . . totam ex integro terram suam cultam et incultam quam habebat vel habere debebat per se vel per aliam personam in loco illo qui vocatur Comas. . . . Et pro predictis donis predictus abbas donavit ex caritate predicte donatrici X et VIII sol. morl."

[16] *Cart. de Bonnevaux*, no. 277 (n.d.), and no. 291 (1169).

St. Martin of Colons, St. John of Celsan, and St. Peter of Beceria.[17] Similarly, the parishes of St. Just of Durban and of la Rochelongue, as well as the church of Montlaurès and the monastery of Montveyre, were acquired by Fontfroide.[18] Dedications of churches coming into Cistercian hands tended to be to saints popular in earlier centuries, like St. Félix, Ste. Foi, or Ste. Eulalie.[19] An expert on parish dedications might be able to estimate the age of parishes like those dedicated to the popular early medieval solider and abbot, Saint Martin of Tours. In any case, parishes which were new in the twelfth century would more often have been dedicated to the Virgin Mary, and even newly established parishes would still be indicative of previous settlement.[20] However, the fact that the tithes in such parishes had already been alienated to laymen and that even the churches themselves were often in lay hands indicates the antiquity of these parishes, as well as the long-settled condition of the land in those parishes which were coming into Cistercian hands. Such parishes had been producing tithes for the support of parish churches and their parish priests for a considerable time, and were not in areas of new settlement.[21]

In areas where granges were founded, Cistercian abbeys often ended by acquiring both tithes and parish churches. Such acquisition of church property which was more properly held by bishops and parish priests, although strictly speaking in contradiction to the order's early ideals, should not be considered a measure of Cistercian decadence. It represents, instead, the need to compromise with the conditions of land-tenure in new regions into which monks and reform ideals came, and the search for unencumbered lands in southern France on which to practice agriculture. Such unencumbered lands could not be obtained through simple purchase or donation from a single previous owner, but had to be "created" through multiple acquisitions from numerous owners including owners of tithes and churches.

A good example of how charters concerning tithes provide evidence of previous settlement and of peasants resident on land coming into Cistercian

---

[17] Montpellier, A. D. Hérault, film, "Cart. de Valmagne," I, fols. 58v (1146), 58r (1158-73), 80r (1168).

[18] Carcassonne, A. D. Aude H 211 "Inv. de Fonfroide," fols. 1r, 61r, 61v, 62r.

[19] Examples of parish dedications in the region of Gimont, for example, include place names mentioned in the cartulary such as: Saint-Antonin, Saint-Aubin, Saint-Caprais, Saint-Clar, Sainte-Christie, Sainte-Foy-de-Peyrolières, Sainte-Marguerite, Saint-Gaudens, Saint-Geni, Saint-Georges, Saint-Germier, Saint-Guiraud, Saint-Jean, Saint-Jean-de-las-Monges, Saint-Julien d'Aiguebelle, Saint-Justin, Saint-Laurent, Saint-Lederur, Saint-Martin, Saint Martin-de-Garbic, etc., as well as Sainte-Marie, Sainte-Marie-la-Garnison, and Sainte-Marie-la-Grasse—*Cart. de Gimont*, 487-89.

[20] See Charles Higounet, "Hagiotoponymie et histoire. Sainte Eulalie dans la toponymie de la France," *Actes et Mémoires du Cinquième Congrès international de Sciences Onomastiques* 1: 105-13, and idem, "Les hommes, la vigne et les églises romanes du Bordelais et du Bazadais," *Revue d'histoire de Bordeaux* 1 (1952): 107-11.

[21] See Constance H. Berman, "Cistercian Development and the Order's Acquisition of Churches and Tithes in Southern France," *Revue bénédictine* 91 (1981): 193-203.

ownership is found in the arbitrated settlement of a dispute over tithes between the Cistercian abbey of Bonneval and the owners of the parish church of Corronzanges in the Rouergue. This document, in which the tithes pertaining to certain *mansi* in the parish of Corronzanges were conceded to Bonneval, shows that by 1176 Bonneval was already in possession of a number of such *mansi* in that parish.[22] Moreover, the charter reveals that the owners of the church of Corronzanges and the monks had agreed that if in the future additional peasants (*coloni*) from that parish were to cede their holdings to Bonneval, then the monks of that abbey would compensate the church of Corronzanges by paying it ten *solidi* for every hundred *solidi* expended for such land acquisitions. The significance of this act is that it describes a parish in which peasants were living and had lived, in which peasants themselves had conveyed rights of cultivation to the monks in return for cash, and in which tithes were established claims on the land. This was not an area of "new" land or abandoned land or wilderness of any kind. Peasant's rights were so well established that they were heritable, and land did not have "noval" tithes on it.[23] The Cistercians of Bonneval acquired such a territory not by reclamation, but by gift or purchase on the land market.

A third index of previously established cultivation on land conveyed to the Cistercians in southern France is found in references to peasant "proprietors" on land transferred by charter to the order. In addition to references to peasant settlement found in formulaic descriptions as already cited, this evidence includes references to peasants mentioned in order to identify land conveyed, those to peasants whose rights to land were so secure that they figure as "donors" or "vendors" of their hereditary rights, lists of rents owed by peasants who were tenants, and charters which specifically mention the removal or resettlement of peasants living on lands conveyed to the order.

A number of twelfth-century charters for the Rouergue in particular mention conveyance of a peasant's right to cultivate a holding using an explicit term for such rights: *pagesia*.[24] Such concessions of *pagesia* were generally made for such traditional holdings as *mansi, caputmansi,* and *appendariae*.[25] In a few cases *pagesia* rights appear to have been under the control of a lord who promised Bonnecombe, for instance, that the monks might have *pagesia* rights from any of the men under his authority, or have

---

[22] *Cart. de Bonneval*, no. 23 (1176).

[23] Noval tithes were on lands never before subject to tithe and attached to no previous tithe owner. See Constable, *Monastic Tithes*, 105–106; the charters for Hauterive contain some interesting examples of lands on which tithes were noval, see *Liber donationum Altaeripae: Cartulaire de l'abbaye cistercienne d'Hauterive (XIIe–XIIIe siècles)*, ed. Ernst Tremp, trans. Isabelle Bissegger-Garin (Lausanne: Société, 1984), no. 163, for example.

[24] Niermeyer, *Lexicon*, 752.

[25] Rodez, Soc. des lettres, Bonnecombe, "Cart. de Magrin," nos. 1 (1167), 2 (1170), 5 (1172), 13 (1177), 32 (1183), etc. and *Cart. de Bonnecombe*, nos. 258 C (1178), 264 H (1183), and 268 C (1187); see quotation in n. 27 below.

such rights in any place where he had granted rights of *fevum* to the monks.²⁶ More often conveyances mentioning *pagesia* have as principals the peasants themselves. An example is the gift to the monks of Bonnecombe by Bernard of Bez and Bernard his son of their *pagesia* in the *mansus* of Bez and anything else they had there; in return, Bernard of Bez senior was to be received as a *conversus* at Bonnecombe if his wife absolved him of their marriage vows and if he came without any fuss or clamor.²⁷ In a few cases from the Rouergue, a charter mentions rights to *fevum* and to *pagesia* together. *Fevum* throughout the Midi often meant a tenancy rather than "fief" in the sense of a hereditary knight's tenure granted in return for military service, and documents from the south mentioning concession of *fevum* rights must often have concerned rights to cultivate a tenancy; thus, *fevum* was probably often identical to *pagesia* in these charters.²⁸ Similarly, Silvanès's documents indicate that concessions of *beneficium* were also often identical to conveyances of *pagesia* or rights to cultivate; in at least one case, the Silvanès charters show *beneficium* over land conceded by individual men (presumably of the peasant class) who were also absolved of *servitia* and *usatica* on that land, and were then admitted to the abbey as lay-brothers or *conversi*.²⁹

Occasionally charters record the enfranchisement of individuals who simultaneously conceded their previous tenancies to the Cistercians; these too demonstrate that Cistercians were acquiring land already cultivated. There are several such examples in the Grandselve cartularies. Bernard and Raymond of St. Germain, for instance, were granted liberty by Lady Assault in return for their concession to Grandselve of rights over land which they had cultivated; Bernard of las Bordas was similarly freed after his land was given to that abbey.³⁰ Such enfranchisement is also indicated

---

²⁶ Rodez, Soc. des lettres, Bonnecombe, "Cart. de Moncan," no. 40 (1172).
²⁷ *Cart. de Bonnecombe*, no. 264 H (1183): "M.C.LXXX.III. Ego Bernardus de Becio et ego Bernardus, filius ejus, nos per nos . . . donamus et concedimus Deo et Beate Marie Bonnecumbe et Randulfo, abbati et vobis aliis fratribus ejusdem loci . . . , pagesiam mansi de Becio et quicquid aliud ibi habemus vel habere debemus vel aliquis homo vel femina per nos vel de nobis, ut libere et quiete habeatis et possideatis. . . . Et nos supradicti fratres te Bernardum de Becio seniorem propter hoc debemus suscipere in conversum, mox ut uxor tua te absolverit, et ad nos venire poteris sine ullo clamore."
²⁸ Hubert Richardot, "Le fief roturier à Toulouse aux XIIᵉ et XIIIᵉ siècles," *Revue historique du Droit* 40 (1935): 307-59.
²⁹ *Cart. de Silvanès*, nos. 220 (1162), 319 (1167), and 224 (1163); the last says, "M.C.LXIII., ego Ugo de Serrutio et uxor mea Guillerma et infantes nostri, scilicet Ugo et alii . . . , donamus et laudamus et titulo donationis cum hac carta tradimus monasterio beate Marie de Salvanesc et tibi Pontio, abbati, . . . totum quod habemus vel habere debemus in illis faissis quas habuisitis a Guillermo Ademari de Montealegre, scilicet fevum et beneficium; et solvimus et dimittimus omnem clamationem et querimoniam quam faciebamus in manso de Peritio et in caputmansum quam tenuit Raimundus Textor; et si quid juris habemus vel habere debemus in territorio de Promilach, totum donamus et laudamus; et dedistis nobis L vacivos de caritate, ut omnia predicta firma . . . et sciendum est quod me predictum Ugonem debetis recipere pro converso, si secundum ordinem vestrum ad vos venire potero. Testes hujus rei sunt."
³⁰ Paris, B. N. Latin MS 9994, Grandselve, no. 650 (1163), and Latin MS 11009, Grandselve, no. 80 (1172).

in acts for Berdoues, Bonnefont, and Candeil.[31] Explicit transfer of lordship over peasants along with the land which they held, also implies previously tilled holdings. For example, when Grandselve acquired the *villa* of Toalis, Vidal of Bairis gave the abbot all his claims to the men and women who were tenants there.[32] Similarly, the rare conveyance of land specifically designated as being held by an allodial peasant—one without any lord— also indicates previously settled land. In one charter, Bruno of Lacome, son of Vidal of Lacome, gave to Berdoues the *casal* of Lacome which he held at Durfort, indicating to the monks that it was free of services owed to any lord, and free and immune from any lord; in return he was promised admission at Berdoues if he wished to enter.[33] A document in Bonnecombe's Bernac cartulary lists by name the men and women with their progeny who belonged to the half of the *villa* of Brins which had been conveyed to Bonnecombe,[34] and in Provence, peasant-tenants were allowed to remain under the authority of the Cistercians of Léoncel in a *villa* which the monks has acquired jointly with the church of St. Ruf.[35] Although examples such as these in which lordship over peasants is explicitly discussed are isolated, the general tenor of the charters suggests that a more widespread transfer of peasants or lordship was underway, although it may often have gone undocumented.

A long history of settlement on lands acquired by the Cistercians in southern France is also indicated by the conveyance of fragmented holdings to the monks. An extreme example of such acquisitions of fragmented rights and claims by the Cistercians is provided in documents for the *villa* or village of Magrin acquired by the monks of Bonnecombe in the Rouergue. This village, located not more than a kilometer from the abbey of Bonnecombe, contained as many as twenty or thirty *mansi:* those of la Carpentaire, given by Peter Carpenter and Bernard Carpenter, that of Maselz given by Bernard of Maselz, that of Fraisse and of Ugonenc given by Hugh of Gaillac and his associates, that of Cairo given by Ademar of Cairo on his deathbed,

---

[31] Paris, B. N., Coll. Doat, vol. 115, Candeil, no. 123 (1209) (fol. 19r ff.); *Cart. de Berdoues,* no. 154 (n.d.), and *Rec. de Bonnefont,* nos. 221 (1205) and 282 (1231).

[32] Paris, B. N., Latin MS 11008, Grandselve, fol. 104v–105r (1188) and Latin MS 9994, Grandselve, fol. 217b (1196); *Cart. de Berdoues* no. 534 (d.n.).

[33] *Cart. de Berdoues,* no. 457 (1208): "Sciendum est quod Brunus de Lacoma dictus filius Vitalis de Lacoma, . . . in elemosina et in caritate dedit Deo et beate Marie et Gillelmo, abbati Berdonarum . . . totum ex integro casale de Lacoma, quod est a Durfort cum omnibus pertinentiis suis, scilicet omnes terras cultas et incultas, quas ibi habebat vel habere debebat, per se vel per aliam personam, pascua et erbagges et omnem espleitam, liberum introitum et exitum ab omni servitute alicujus domini immunem et liberum, ut predicti fratres Berdonarum habeant et possideant predictum casale sine omni sua et suorum contradictione in perpertuum. Et debet inde facere bonam et firmam garentiam de omnibus amparatoribus predictis fratribus Berdonarum, tali vero pacto quod si ejus nepos Laurentius, forte vellet manere in predicto casale, posset recuperare et habere medietatem predicte terre. Et pro hoc dono predictus Brunus receptus est in capitulo Berdonarum et debent eum recipere, si ad religionem secundum formam ordinis venire voluerit. . . . M.CC.VIII."

[34] Rodez, A. D. Aveyron, 2H Bonnecombe, "Cart. de Bernac," no. 142 (1224).

[35] *Cart. de Léoncel,* no. 11 (1163–65).

and so on.[36] Many of these holdings in the *villa* of Magrin had been subdivided into half and quarter *mansi*, and all had multiple levels of ownership over them at the time of Cistercian acquisition. Such fragmentation of tenancy units meant that rights to cultivate (*pagesia*) were held in eighths for a single *mansus*, or tithes were divided into twelfths, or lordship was held in eighths by the various heirs of the original castle-holders; fragmentation of the levels of ownership meant that allodial rights were in different hands from lordship, and so on.[37] The consolidation of subdivided holdings in the village of Magrin itself, however, was not in itself sufficient to create a Cistercian grange in this instance. Settlement and perhaps even crowding had gone beyond the village. In addition to the fragmented holdings at Magrin, the monks also acquired the isolated *mansi* located in the interstices between Magrin and other villages along the Viaur valley where Bonnecombe was sited. Conveyances of such scattered claims over land and its produce came to Bonnecombe from a variety of individuals with claims to the land which would eventually make up the grange of Magrin.[38] Often such rights were so dispersed that they had become virtually valueless to their owners. As discussed in the next chapter, such rights became profitable to their new owners only after their consolidation and reorganization by the monks.

Elsewhere in the Rouergue, fragmentation was probably not as advanced as at Magrin, or at least the Cistercians were not involved in quite such a striking reorganization of fragmented holdings as that carried out by Bonnecombe. Among conveyances to Silvanès in the southern Rouergue, half-*mansi* were quite frequent, but it is unusual to find *mansi* divided into quarters.[39] There, too, however, as at Magrin, holdings which had been dispersed among many owners may well have been nearly profitless until they were reorganized. For instance, many charters show small strips of

---

[36] Rodez, Soc. des lettres, Bonnecombe, "Cart. de Magrin," nos. 1 (1167), 2 (1157), 5 (1172), 8 (1174), 49 (1175), etc.

[37] Ibid., no. 2 (1170): "Bernardus delz Maselz dat suam partem pagesiam delz Maselz," no. 13 (1177): "Petrus delz Teillz et filius eius dant quartam partem pagesiam mansi delz Bordelz," no. 14 (1178): "Ugo comes et uxor eius dant alodium mansi delz Maselz et delz Bordelz," no. 39 (1183): "Bernardus d'Arpaio dat alodium delz Maselz"; *Cart. de Bonnecombe* no. 25 (1182): "Rigaldus Amblartz dat partem vicarie de Bordelz"; other examples include ibid., no. 33 (1183) one eighth of *pagesia* of *mansus* and *caputmansus* of Bargas, and *Cart de Bonnecombe*, no. 259 B (1179): "M.C.LXX.IX. Ego Bernardus de La Roca, per me . . . , dono et concedo sine omni retentione Deo et Beate Marie Bonnecumbe et Ugoni abbati, . . . medium feudum et totum quod habeo et habere debeo in manso et in capmanso de Macellis, similiter et in manso de Bordellz et medium capmansum de Papa Gall, ut libere et quiete habeatis et possideatis perpetuo jure; et vos dedistis michi caritative quadringentos LXX solidos, et si prefatus honor amplius valet, dedi domui Bonecumbe et habitatoribus ejus pro amore Dei et redemptione peccatorum meorum.—Hanc donacionem suprascriptum laudavit et concessit Petronilla, uxor predicti B., et G. de La Roca, filius ejus."

[38] Rodez, Soc. des lettres, Bonnecombe, "Cart. de Magrin," nos. 92 (1201), 91 (1203), 85 (1200), and 14 (1178); see below, chapter 4, pp. 70–71, and Berman, "Administrative Records," 208–11.

[39] For example, *Cart. de Silvanès*, no. 358 (1138), "Totum hoc quod habebam in manso Felgairetas qui est ad Fontemfrigidum, hoc est fevi medietatem."

vines or fields being reassembled by the monks of Silvanès.[40] In some cases, such consolidation included the reassembling of a preexisting *villa*, like that of Laurs which was incorporated into Silvanès's grange of Promillac. In this case, too, the grange property eventually extended considerably beyond the earlier *villa*.[41]

Evidence for such fragmentation among holdings coming to the Cistercians in other parts of the south did not always comprise references to halves and quarters of *mansi* or tiny strips and parcels of land. In many cases, simply the presence of large numbers of individuals conveying rights over a single farm or *villa* to the Cistercians is sufficient to suggest that ownership had become dispersed. Such evidence is also found for abbeys west of the Rouergue, for instance, in the Grandselve cartularies. There, multiple conveyances by a large number of individuals, of land located in the same place or territory imply fragmentation. Along the Save River south of Grandselve, for example, that abbey acquired rights for its grange or cattle station of Larra in a variety of holdings at such places as la Foisse, the adjoining Boseville, and at Goiac, Loubeville, Gaugiac, and Tirapel; some of these conveyances were of rights in individual *casaux*, some of them of rights to lands within an entire territory.[42] Similarly, at la Terride on the Gimone River, Grandselve acquired by multiple conveyances from a variety of donors the vaguely described rights to land in the territories or parishes of the churches of Usac, St. George, la Terride and St. Peter of Vinsac, as well as in the territories of Hauteville, Albarelle, Masdoat, and Solbario.[43] Such conveyances of land came to the order from a variety of owners, all of whom had vague rights in some territory.

Such a history of occupation and settlement is also implied by dispersal of rights in another sense than strict fragmentation of holdings, that is, the layering of level upon level of ownership to the land or of claims to its products. This second type of diversification of ownership included not only the usual medieval division of property ownership into lordship and usufruct, but additional division of ownership rights over a specific piece of land, however undivided that unit of land itself might be; tithe-holders,

---

[40] Ibid., nos. 452–56 (1163–64). Similar consolidation of vineyards is found for Grandselve; see Toulouse, A. D. Haute-Garonne, 108H58 Grandselve (cartulary rolls), and 108H, Grandselve, liasses 1–12.

[41] Consolidation at the grange of Promillac required seventy-four acts; see below chapter 3, p. 49.

[42] Paris, B. N., Latin MS 11008, Grandselve, fols. 125–80, contains more than thirty conveyances concerning a single place called Foissa. In the same cartulary, many others concern the grange of Vieilleaigue. Boseville, la Foisse, and the parish of the church of St. Severin were frequently associated in a single act; for example, ibid., no. 93X (1167). Goiac was a parish in its own right; see ibid., no. 18X (1161). Honors in the territory of Loubeville were located between the honors of Goiac and of Boseville; see ibid., no. 22X (1162), the honors of Gaugi and Tirapel were also adjacent to those territories; ibid., no. 74X (1165), and no. 79X (1165). The *villa* of Goiac was on the Save River; ibid., no. 241X (1180), etc.

[43] Paris, A. N. L1009 bis, Grandselve nos. 129 (ca. 1150), 46 (1161), 45 (1162), etc. Many related acts are found in Paris, B. N. Latin MS 11009, Grandselve.

tithe-collectors and other middle-men, the holders of vicarial rights, owners of *allodium* as opposed to *dominium*, owners of *fevum* as opposed to those of *beneficium*, and so on, as well as lords and peasants, all might have had claims to the same piece of land.[44] In the most extreme cases, as at the village of Magrin discussed above, there were even lenders holding mortgages over certain rights to the land.[45] Generally, however, the two types of dispersal of land were virtually indistinguishable, with the fragmentation of ownership rights among many owners often accompanied by the actual subdivision of holdings.

With such dispersion of rights, of course, came diminishing returns to agriculture. This suggests that in many cases where fragmented holdings in southern France were given or sold to the Cistercians, those holdings were the equivalent of the marginal lands so often given to monastic groups. Moreover, it is likely that such lands had been conveyed to the monks because it had become very difficult to make them profitable.[46] With so many owners and claimants such lands could only be converted into workable and profitable estates because religious groups like the Cistercians had the cash available to repurchase rights from a host of "owners" and intermediaries who had claims to it. This means that the expenditure of large amounts of cash in numerous transactions for many relatively small holdings is itself an index of the settled condition of land which came into Cistercian ownership. The many conveyances necessary and the large expenditures needed for such acquisition (as indicated in Table 2) confirm that the early Cistercian monks did not create their estates by laboring on the "new" lands or on the "internal frontiers" of the twelfth-century Midi. The documents suggest that most land was occupied, and was certainly not readily available and free to anyone in southern France who desired to clear it for his own use. Cistercian expenditures for particular parcels might be large or small, and acquisitions by any abbey or for any particular property few or many, but purchasing and consolidation of land is suggested in many southern French Cistercian charters.[47]

A final body of evidence showing that the Cistercians acquired land which had a long history of cultivation is found in charter references to "development" or "improvement" on land coming into the order's patrimony—in descriptions of existing villages, houses, barns, vineyards, gar-

---

[44] In Bonnecombe's acquisitions at Magrin, for example, the counts of Rodez held allodial rights in several *mansi* at Magrin (see n. 37, above), the lords of Calmont held *dominium* in the *villa* of Magrin, and the counts of Toulouse-Rouergue held judicial rights nearby which were ceded eventually to Bonnecombe; see Rodez, Soc. des lettres, Bonnecombe, "Cart. de Magrin," nos. 14 (1178) and 85 (1200), and *Cart. de Bonnecombe*, no. 51 (1244).

[45] Generally, Cistercians in southern France did not become involved in lending against mortgages, although unredeemed lands under mortgage came into their hands; for exceptions, see Constance H. Berman, "Land Acquisition and the Use of the Mortgage Contract by the Cistercians of Berdoues," *Speculum* 57 (1982): 250–66.

[46] For amounts spent by the largest abbeys of the region, see Table 2.

[47] The conveyance of marginal lands to Cistercians elsewhere has been best documented for Britain. See Donkin, *Cistercians*, 103–34, and Hill, *English Cistercian Monasteries*, 53 ff.

dens, and orchards, as well as in references to the existing tithes and churches already mentioned. Previous occupation is obvious in conveyances of such already developed holdings as the vineyards at Bougaunes conveyed to Bonnecombe, other vineyards granted elsewhere to the order, and the rights to olive orchards and olive trees at St. Cyprien collected by Grandselve.[48] Although one of the economic contributions made by Cistercians in southern France was probably the additional improvement of land, a certain amount of development had already taken place before Cistercian acquisition; for instance, many mills, weirs, and associated mill-ponds had been constructed long before they became the property of the monks, although they were often in a state of disrepair when conveyed to the order, as discussed further in chapter 4. Similarly, even in areas in which clearance had taken place only fairly recently, often done by the immediate predecessors of the order, most land had nonetheless already been cleared. Many properties had farm buildings, churches, planted vines, and orchards already in place. Such improvements would not have been found on a frontier, any more than would the existing peasants, or traditional tenurial names, and so on.

The above discussion provides examples of the various types of evidence contained in a much larger body of documentation confirming that the creation of new arable lands in southern France did not coincide with the coming of the Cistercians. Whatever clearance, reclamation, or reinstitution of cultivation had occurred after ca. 950 in that region—in wilderness areas, in marshes, or on lands abandoned during Saracen raids—seems to have been done well before the Cistercians arrived, or only in the late thirteenth century under extreme population pressure.[49] This is true even with regard to drainage in coastal areas like the Rhône delta or Camargue. There the Cistercians had houses at Sauveréal and at Franquevaux; little can be said about Sauveréal, but Franquevaux was built on the edge of the salt marshes in the region between St. Gilles and what would become Aigues-Mortes,

---

[48] For Bougaunes, see Rodez, Soc. des lettres, "Cart. d'Iz et Bougaunes," passim. Other vineyards are conveyed in Rodez, A. D. Aveyron Bonnecombe, "Cart. de Bernac," nos. 34 (1195), 39 (1196), etc., or Paris, B. N., Coll. Doat, vol. 70, Villelongue, no. 18 (1152) (fol. 36r), no. 24 (1157) (fol. 48r), no. 30 (1165), (fol. 66r), and no. 25 (1158), (fol. 50r); or Paris, B. N., Coll. Doat, vol. 83, Boulbonne, nos. 146 (1231) (fols. 336v ff.) and 86 (1195) (fol. 174v); or Paris, B. N., Latin MS 9994, Grandselve, no. 752 (1157): a list of eighty plots with olives and fifty places where rents on olive trees were held by Grandselve.

[49] This is also my finding from additional work on all monastic charters (not only Cistercian) for the region of the Carmargue; it will be discussed in a forthcoming article, "Pre-Cistercian Settlement and Reclamation in the Camargue." There is considerable recent evidence that there was much less clearance activity by Cistercians elsewhere than had previously been assumed; see references in chapter 1, n. 29. André Chédeville, in *Chartres et ses campagnes (XI*-XIII* siècles)* (Paris: Klincksiek, 1973), 113–14, refers to acts giving Cistercians permission to clear land in northern France, but he shows in at least one case, involving the Cistercian monks of Les Vaux-de-Cernay to whom permission to clear land was given in the midthirteenth century, that much less clearance was actually done there than had been allowed. See also my recent findings on clearance activities by peasants under the control of Cistercian women's houses in the region of Chartres, in "Examples," forthcoming.

near what is even today apparently uninhabited territory. Its arable lands, however, as is very clear from surviving documents, were almost invariably found not in the Camargue marshes but farther inland on higher ground, and they had almost all been planted with vines or had begun producing cereals before monastic acquisition.[50] Everything points to the conclusion that even along coastal marshes, Cistercians did not initiate reclamation.

Moreover, there is considerable evidence against any contention that the order's acquisition of lands in areas of previously settled agriculture was accompanied by significant frontier activities which simply have escaped documentation. Although there had been times in the Middle Ages when forests were infested by a great variety of squatters—thieves, charcoal-makers, or peasants working a bit of extra land without authority—that situation had been the result of weakness on the part of the lords who could not enforce their rights, and not because the forest was considered open and free to anyone who undertook such activities.[51] By the twelfth century, when lords became stronger, forests had already shrunk, had become valued holdings, and were no longer free to those who desired to uproot trees and create new agricultural land for themselves by squatting. Permission had to be received from owners for any clearance activities and, indeed, grants of such permission allowing clearance in southern France by monastic corporations and their tenants are recorded in charters for religious houses of the eleventh century.[52]

It must be assumed that if such permission had been granted to the Cistercians, it would have been noted in their records. Contracts giving permission to clear land are almost never found in these Cistercian cartularies, and, given the overwhelming evidence of land acquisition by other means, it must be concluded that reclamation was not a source of Cistercian land in southern France, except in very rare cases. Indeed, Cistercian documents which give even a vague suggestion of possible reclamation by the order are so infrequent that all may easily be cited here. Alongside thousands of documents recording conveyance of previously cleared and previously cultivated lands to Cistercian houses in southern France, only a handful of acts, less than *twenty* for all the abbeys in the entire region, could possibly be interpreted as giving permission for either clearance or

---

[50] Nîmes, A. D. Gard, Franquevaux H 40 no. 1 (1200), *mansus* of Guraldinus given by Bremund William of Sommières, and H 42, no. 1 (1201), the conveyance by Anastasia, widow of a certain Pons to Franquevaux of two pieces of land in the territory of la Rovière in the parish of Belvézins.

[51] Duby, *Rural Economy*, 143–45.

[52] For earlier houses, see Higounet, "L'occupation du sol," 227–54; Toulouse, A. D. Haute-Garonne 205H Lespinasse liasse 12; and *Charters of St. Furzy of Peronne*, ed. William Mendel Newman with Mary A. Rouse, preface by John F. Benton (Cambridge, Mass.: Medieval Academy of America, 1977), no. 11 (1119), which shows arable land, two new settlers, and woods given to the monks; this suggests quite recent settlement. Ibid., no. 26 (1162) concerns noval tithes. I would question, however, whether the *terra culta et inculta* in that region, any more than in southern France, indicates virgin territory.

drainage of land by the order. These rare documents, many of them suggestive of reclamation rather than clear evidence of it, or implying not Cistercian reclamation but that by their immediate predecessors, confirm that if significant pioneering activities had been undertaken by the Cistercians, they would have been recorded in the cartularies.

For the Cistercian abbeys of Bonnevaux and Léoncel east of the Rhône, several charters grant permission to irrigate meadows for the production of hay, but these are not so much an instance of reclamation as of capital improvement; Bonnevaux also received one conveyance mentioning land full of briars to convert to meadow which may have actually been waste to be reclaimed, and a concession of land which the monks (?) had uprooted at Chalvas.[53] In the region east of the Rhône, too, the monks of Franquevaux received a concession of cultivated land at la Roquette and rights to uncultivated land there if they cleared it.[54] On the Languedoc coast, however, although land in only recently developed *villeneuves* sometimes came into the hands of the abbey of Franquevaux, there are no instances of reclamation along the lagoons by that abbey or by Valmagne or Fontfroide.[55] A number of single holdings, described as *heremus* or *herema*, were conveyed to the monks of Valmagne, but these were always in the vicinity of previously cultivated lands.[56] Similarly rights were given to the monks of Valmagne to construct a grange or another abbey in the forest of Burau; in a papal confirmation dated 1185 a grange at Burau is mentioned, but there is no indication that this forest was not already a developed area.[57] One contract also conveyed to that abbey a pond which could be drained for cultivation or used for a milling complex, but it is clear from other documents that it was a mill that was constructed there; in a similar contract Valmagne was given property adjoining a salt-marsh, but this was obviously

---

[53] *Cart. de Léoncel*, no. 169 (1251), and *Cart. de Bonnevaux*, nos. 212, 271, 287, 288 and 204 (n.d.); to quote from the last: "Ismio Ruvoiria dedit supradictis fratribus quoddam duimetum (sic) apud Aquam Bellam, ad pratum faciendum," and *Cart. de Bonnevaux*, ed. anon., no. 91 (n.d.): "de ruptis quas fecerant in Chalvas."

[54] Nîmes, A. D. Gard, H33 Franquevaux, "Inv." fol. 178v, no. 2 (1157).

[55] There is one charter dated 1168 for Franquevaux in which Raymond V of Toulouse confirmed the conveyance made to the monks by William Hipolytus and his wife of a land or marsh (*paluda*) which was already adjoined on all sides by cultivated holdings; later in the same document it is referred to simply as "land": Nîmes A. D. Gard, H43 Franquevaux, no. 1 (1168). While this may be evidence of fairly recent drainage of marshland, it does not seem to have been carried out by the Cistercians. On new settlements or *villeneuves* acquired by Franquevaux, see, for example, Nîmes A. D. Gard, H33 "Inv. de Franquevaux," fol. 153v, no. 3 (n.d.).

[56] Montpellier, A. D. Hérault, film, "Cart. de Valmagne," vol. 1, fol. 133v (1195), "unum heremum in termino de Vairaci," which was in the midst of the *villa* of Vairac, one of the most settled areas acquired by Valmagne, if one can judge from the prices for land there and the numbers of charters involved; other *heremi* are mentioned ibid., vol. I, fols. 69r, (1188), 94 r (1182), vol. II, 109v (1198).

[57] Montpellier, A. D. Hérault, film "Cart. de Valmagne," vol. I, fol. 6v (1185), is the papal bull; charters are found in vol. 1, fol. 83v ff. The fact that a parish, that of St. John of Fraissenet, is in that forest, suggests that it was inhabited already.

for salt-production, rather than for reclamation and conversion into arable land.⁵⁸

Charters for the Cistercian house of Villelongue near Carcassonne include a single act allowing brush to be cut along the Alzonne River; although nothing is said about cultivation, this may be an authentic example of permission to reclaim land and create cultivable holdings.⁵⁹ Among charters for the abbey of Silvanès in the southern Rouergue there is an early charter referring to a *mansus hereminius* [sic] *et condrictus* where the hermits who preceded the Cistercians at Silvanès may have constructed their house.⁶⁰ Such a phrase is, however, the equivalent of "in terris cultis et incultis," indicating that both cultivated and fallow parts of settled territories were given to those men. Silvanès also received permission to construct dams near a milling complex in order to irrigate meadows.⁶¹ No evidence of clearance by the monks can be found among its charters.

North of Toulouse, the monks of Grandselve received land in mortgage from which brush could be cleared to reduce it to cultivation; however, whether they undertook this project is not clear.⁶² For that region there

---

⁵⁸ Ibid., vol. I, fol. 108v (1182): "M.C.L.XXX.II . . . Ego Bernardus Ermengaudi et ego Raimundus frater eius et ego Gaucelmus frater eius donamus, laudamus et concedimus domino Deo et Beate Marie Vallismagne . . . quantum nos habemus in tenemus de Bernardo de Capraria in toto stagno de Tortoreria quod est in terminio de Vairaco, . . . possitis aquas dirivare et stagnum desiccare et sic ad ultimum culturam redigere; si autem molendinos construere in predicto stagno vel de ipsa aqua . . . aut arbores plantare sive pisces nutrare. . . . concedimus." Their sister received 300 sol. for this concession and they give a guarantee of their rights over an adjoining *condamina*. It appears from other charters that Valmagne continued to have a mill there rather than draining this pond, see ibid., vol. I, fols. 100v–10r. The other pond is mentioned ibid., vol. I, (fol. 140v) (1176): "Unum campum ad salinarias faciendas . . . [adjoining] ex altano cum stagno qui vocatur Taurinus." This document is discussed by André Dupont, "L'exploitation du sel sur les étangs de Languedoc, IXᵉ–XIIIᵉ siècles," *Annales du Midi* 70 (1958): 14.
⁵⁹ Paris, B. N. Coll. Doat, vol. 70, Villelongue, no. 52 (1202) fol. 114r ff.; "Donamus totas nostras bodigas, quicquid habemus . . . ad Alsonam ripariam, et rumpatis totas dictas bodigas et . . . laboratis."
⁶⁰ *Cart. de Silvanès* 11 (1132): "Totum hoc quod habemus et habere debemus ad Terundum, scilicet medietatem mansum herminium et condrictum."
⁶¹ Ibid., no. 210 (1159), "Paxeriam ad prata facienda in Promilac."
⁶² Paris, B. N. Latin MS 11011, Grandselve, fol. 171r (1182): "M.C.LXXX.II. . . . Ego Sasneus qui fui uxor quondam Cenabrum de Canzas, . . . mitto in pignore tibi Willelmo abbati Grandissilve et fratribus eiusdem loci . . . propter C. sol. Tol. totum honorem meum cultum et incultum cum omnibus suis pertinentiis, quem habeo et habere debeo sive homo vel femina habet a me et habere debet, quocumque modo in territorio Sancti Laurencii: scilicet infra nausam Sancti Laurentii et nausam de Verned et viam que descendit a Fonte de Cumbas usque ad predictam nausam del Verned et terram vostram quam habetis a parte castri de Assina, . . . ut tam diu habeatis et teneatis libere et quiete donec predictos C. sol. Tol. vobis reddam. Et tamen retineo mihi in istud pignus tantum anno presenti: scilicet omnes redditus fructuum et segetum qui de isto honore exierint usque ad futurum festum Sancti Saturnini quod erit presenti anno. Iam vos postea sic in potestate vostra omnes predictus honor ad omnes voluntates vostras faciendas quousque reddam vobis iamdictos C. sol., si tamen usque ad predictum festum Sancti Saturnini non reddidero vobis. Si autem bosquos et bartas et bozigas que in isto honore sunt, succideritis et redigeritis ad culturam, concedo vobis ut tamdiu ipsum honorem quem ad cultum redigeritis habeatis et teneatis donec tres segetes planas? mihi habeatis. Mando etiam vobis ac permitto quod de omnis predicto pignore fui vobis

were also a number of conveyances to Grandselve of only recently cleared land which adjoined marshes and swamps (*nausae*).[63] This was one area, among the many examples of previously settled and cultivated land which Grandselve had acquired, where Cistercians may actually have completed reclamation begun by earlier individuals. Even in this case, documents suggest that the bulk of the reclamation was already done, having been completed by the immediate predecessors of the order. Moreover, acquisitions of such holdings in the marshlands along the Garonne comprise only a small part of the total land conveyed to that abbey.[64] For Gascony a single act mentions conveyance of rights to Gimont to clear land (*exheremare*) adjacent to already cultivated territories; for Berdoues, one act specifically forbade the monks to make *novales* and none concern their creation.[65]

Only in the documents for Boulbonne, located in the Ariège south of Toulouse, is there a long series documenting clearance by a Cistercian community, and even there references to grants of explicit permission to cut

---

legitimus guirents de omnibus amparatoribus. Terminus vero solvendi pignoris erit festum Omnium Sanctorum et sic ab anno in anno, expleitis foris. Quem autem tenet de me Arnaldus Sancti Laurencii in omni iamdicto honore, in terris, in decimis, sive per sirventatico, vel alio quocumque modo, ego vobis mando quod sine omni retinimento ipsi Arnaldo faciam vobis concedere et laudare ut libere et quiete habeatis et teneatis quam diu in pignore habueritis. Convenio etiam vobis fratribus ego Sasneus quod pignus istud nullum hominum sive feminam faciam illud solvere ad usus illorum nisi ad meos proprios usus sive labores."

[63] Paris, B. N. Latin MS 11011 Grandselve, fol. 5R (1154): "quicquid habemus et habere debemus in illam terram de la Pradela, que clauditur a nausa ex tribus partibus, ex alia vero parte clauditur a terra Terreni presbiter"; fol. 3r (1150): "totum casalem de la Graveira, cum nemoribus et aquis ingressibus et egressibus et cum omnibus pertinentiis suis, . . . Quod casale a parte orientali confrontat in domo vostra, a parte occidentali in nausa, a parte vero meridiana in terra vostra, et a parte aquilonis similiter"; fol. 6r (1143): "illam terram quam habeo iuxta terram et vineam vostram tam super viam quam sub via et pratum et mediam partem de la nausa"; next charter (1143): "totam illam terram que est iuxta illam terram quam habuistis de me et extenditur usque ad nausam et usque ad pratum Bernardi de Combas fratris mei et usque ad illam terram que est iuxta terram que fuit Willelmi Siguerii"; fol. 9r (1155): "illam terram quam habeo ad pradelas quae afrontat de oriente in nausa, de meridie affrontat in vostro honore quem comparatis de Bernardo de Combas et Ricardo fratris eius, de occidente afrontat in nausa, de septemtrione afrontant in honore de Grandissilva"; and fol. 11v (1155): "terram de Pradela . . . Afrontat autem predicta terra de meridie in terra ipsius Terreni a septemtrione afrontat in terra Willelmi de la Volvena, ab oriente afrontat in nausa, de occidente afronta similiter in nausa."

[64] Only forty to fifty documents in the Grandselve cartulary number 11011, which has 337 documents in all, or less than a thirtieth of the total surviving documents for Grandselve, mention such *nausae*; in most cases these are the boundaries of land conveyed to the monks rather than being what was conveyed. Furthermore, there is abundant evidence in this cartulary that reclamation had been undertaken before the 1140s by Moissac, by the counts of Toulouse, and by numbers of peasants. Whether the Grandselve hermits were also involved is unclear, but the location is a considerable distance across the Garonne River from the final abbey site of Grandselve.

[65] *Cart. de Gimont*, II, no. 186 (1159): "Sciendum est quod Sancius d'Armadanvilla misit in pignus totum hoc quod habebat in tota villa d'Armadanvilla, et totam decimam ejusdem honoris, excepta quadam parva pecia de terra quam tenet Sancius Bernad ante ecclesiam predicte ville, Bernardo abbati pro XX sol. morl. ad XX annos. Postea redimi poterit de Martror in Martror, expletis foris. Si autem predicti habitatores in predicto honore aliquam terram exheremaverint, habeant et teneant ipsam terram exheremtam post pignus solutum, quantam consuetudo est . . . Anno M.C.L.IX," and *Cart. de Berdoues*, no. 411 (1186).

trees, uproot their stumps, and make arable land represent only a small fraction of the total documents surviving, amounting to five or six charters among a total of more than 150 recorded deeds.[66] There too, such reclamation activities were carried out in an area where clearance had been begun earlier by the order's predecessors, a group of hermits in the valley of the Hers River.[67] As in the acquisition of marshes for drainage by Grandselve, or as with the hermits at Silvanès, it is probable that the major work of reclamation was done before Cistercian incorporation; unfortunately, the history of Boulbonne in the 1160s and 1170s is obscure, and it is impossible to date even the incorporation of that house into the Cistercian order with any confidence.

In addition to these few cases in which rights to reclaim were genuinely granted to the order, there is some evidence in surviving charters which can all too easily be misconstrued as denoting clearance by the monks. Although most land given to the Cistercians appears to have had a long history of settlement, occasionally the order acquired land which had only recently been brought under cultivation by the immediate predecessors of the order. As already mentioned, much of the land north of Toulouse adjoining the course of the Garonne and marshes along that river, had only recently been brought under cultivation when Grandselve acquired it. Similarly, some land obtained by the Cistercians at Franquevaux in the Camargue and by Valmagne along the Languedoc coastal plain may have been only recently brought under cultivation, and much of the land granted to Boulbonne may similarly have been newly cleared. In such cases, later observers have sometimes mistakenly attributed such reclamation at those sites to the Cistercians, because that land was relatively new at the time of the order's acquisition.

Whatever tendency there was for Cistercians to acquire such holdings of "new" lands, this does not mean that they had expended the manpower to create those lands—except in rare and limited instances. In particular, certain areas in Gascony, whose clearance historians have attempted to attribute to the early Cistercians solely because the order acquired them, had actually been cleared in the generation or so before Cistercian settlement. In Gascony such relatively new lands were usually designated by the terms *artigas* or *artigue*, and less often *finage* which mean "assart" or "newly cut clearing."[68] Such land, which terminology indicates to have

---

[66] In the forest of Boulbonne, "Aut in rumpendo, aut in laborando, aut in pascendo," Paris, B. N. Coll. Doat, no. 43 (1167), or land in the woods of Boulbonne, "ad rumpendum et ad terram cultam faciendam," ibid., no. 45 (1168); see also nos. 47 (1167) and 50 (1169), conveyances to the priory of Vajal to clear not long before its incorporation by Boulbonne. Boulbonne's documents are inventoried in the *Histoire générale de Languedoc*, ed. Devic and Vaissette, 8: cols. 1883–926.

[67] The relationship between Boulbonne and the earlier hermits is somewhat confused. See the discussion in Berman, "Growth," forthcoming.

[68] Charles Higounet, "Contribution à l'étude de la toponymie du défrichement, les 'artigues' du Bordelais et du Bazadais," *Paysages et Villages*, 105–10.

been recently cleared, was sometimes the core for the first settlement of Cistercian monks in an area. For instance, such an area of *finages* was used for the site of Bonnefont; in other cases, an eremitical settlement site, located on such a newly cleared area became a Cistercian grange, as is the case for the hermitage at Artigues which became a grange belonging to Berdoues when the hermits entered that abbey.[69] In such conveyances of land called *artigue* or of land in place-names compounded with the word—that is, at l'Artigol, Artiganat, Artigabos or Artinals—the use of the term indicates that the land in question had already been cleared at the time of conveyance. The presence of such terms in early Cistercian documents does not mean that the monks who later acquired those lands were involved in its clearance. Those activities had been carried out by their predecessors: the independent hermits and anonymous peasants who had created such "artigues."[70] Moreover, it must be reiterated that land in such "new" areas comprised only a fraction of any Cistercian abbey's land acquisitions in the Midi.

It is possible that such land which had only recently been brought under cultivation came into monastic hands frequently because of problems inherent in its ownership, which made owners very willing to sell it. For instance, although the rich, untapped soils along the middle Garonne provided initially high yields, the threat of flooding may have made lay proprietorship less advantageous as yields returned to more normal levels. In such cases, only a large institution, like Grandselve, which had adequate cash and grain reserves, as well as fields and granges in other types of terrain, could have afforded to cultivate such low-lying areas.

Despite such exceptions noted, the evidence of the surviving records is clear. The overwhelming majority of Cistercian archival documents record the order's acquisition of already developed land with a long history of settlement in southern France. Cistercians were forced to acquire their lands in this way, for by the time the order began to found and incorporate houses in that region, there was little wilderness left. There is also some indication however, that even in areas of southern France where land remained uncleared and unsettled, the monks chose not to initiate reclamation. As is discussed in chapter 3, the order actually may have avoided acquisition of land needing reclamation because of limited manpower, because such uncultivated land as remained was of very poor quality, and start-up costs would have been high, and because unsettled land had no peasants to recruit as *conversi*. Thus, perhaps Cistercian choices about land acquisition in southern France were dictated by more than the availability of land. Whatever the motivation for such choices, the locally generated charters confirm that land acquisition by the order in that region was almost entirely of previously cultivated territory, not "new" lands.

---

[69] *Rec. de Bonnefont*, nos. 1 and 3 (1137–38), and *Cart. de Berdoues*, nos. 173 (1152), and 266 (1155).

[70] This is a major problem in Traissac, "Les abbayes," 146–52.

III. CONSOLIDATION OF RIGHTS AND RECRUITING PERSONNEL

The surviving charters and cartularies of many early Cistercian houses are eloquent witnesses to the monks' care and patience in acquiring land. Although the records of a few of the most successful abbeys were lost in later disasters, there is a strong positive correlation between making of such records and economic strength.[1] Consequently, the story of how Cistercians gathered their landed wealth is primarily the story of Cistercian successes, for missed opportunities and bad management left less trace in the records. That managerial weakness and mistakes sometimes occurred is nonetheless apparent; by the 1170s there were clear distinctions between rich and poor Cistercian houses in the Midi and by the mid-thirteenth century those differences were firmly embedded.[2] The example of the near-failure of Locdieu is perhaps the most striking and well documented, although it was only one of a number of Cistercian abbeys in the region with considerable economic woes. Houses which began in relative poverty had no later opportunity to approach the levels of wealth of those which had been more successful in the twelfth and thirteenth centuries. In particular, hermitages affiliated by the Cistercians into the line of Pontigny in southern France seem to have remained generally small and unimportant in comparison to houses founded or incorporated into the filiations of Cîteaux, Clairvaux, and Morimond. Although in a few cases a rich house was ruined and recovered only slowly, for instance, Belleperche which was sacked during the Wars of Religion, houses which were poor in the thirteenth century never became rich later.[3]

Some of the differences in wealth among Cistercian houses resulted from unequal opportunities to incorporate the assets, personnel, and goods of earlier religious establishments. Initial "foundation" of many Cistercian houses in southern France was by the incorporation of eremitical groups, bringing with them both members and property rights, which gave the "new" Cistercian house an initial advantage in developing its patrimony.

---

[1] The important houses tended to copy charters more often, which contributed to archival survival; the exception is Bonneval, whose archives suffered a direct fire in the seventeenth century, see *Cart. de Bonneval*, intro. by Rigal, p. ix.

[2] See Appendix I for earlier and later assessments, and Tables 2 and 3.

[3] On Locdieu and Silvanès, see chapter 5, pp. 111–13; Belleperche's property and presumably its archives were damaged during the Wars of Religion. See J.-M. Canivez, "Belleperche," *Dictionnaire d'histoire et de géographie ecclésiastique* 8 (1936): cols. 869–71; for others, Berman, "Growth," forthcoming.

In addition to the hermits who provided the core of the Cistercian house of Silvanès in the southern Rouergue,[4] religious settlements had preceded Cistercian foundations elsewhere: at Bonnefont in the Comminges (an earlier Hospitaller effort),[5] and at Boulbonne in the Ariège south of Toulouse, where a hermitage is recorded as late as the 1160s.[6] The house of Berdoues west of Auch in Gascony was also probably founded on the site of an earlier monastery, and it increased its land-holdings and membership even more by the incorporation of a hermitage at Artigues.[7] Similarly, an earlier hermitage at Compagnes was transformed into a Cistercian community there, which was then moved and renamed Villelongue.[8]

In this regard, the western French reform movement under Robert of Arbrissel and Gerald of Salles was quite important to the settlement of the Cistercians in southern France. Valmagne and Ardorel were both independent monasteries associated with Gerald and his group of western French reformed houses before they were incorporated into the Cistercian order.[9] The great Cistercian house of Grandselve also had some connection with those reformers, probably through Philippa of Toulouse, who had known Robert.[10] That there were hermits at Grandselve is without doubt. Surviving manuscripts of that abbey include an early twelfth-century letter to the king of England from the bishop of Toulouse commending the "holy men living in the forest of Grandselve."[11] Grandselve was already a thriving foundation in the 1140s when its properties and membership were incorporated by Clairvaux at the time of Bernard's preaching mission to Toulouse.[12] Similarly, Grandselve's daughter-abbey, Fontfroide, had earlier been a Benedictine house, probably of eremitical origin, founded in the Languedoc in the late eleventh century. By the time of Cistercian incor-

---

[4] Cart. de Silvanès, nos. 8, 33, and 35 (1135), and no. 470 (chronicle).

[5] Rec. de Bonnefont, nos. 1–3 (1137–38), and pp. 19–20.

[6] Paris, B. N. Coll. Doat, vol. 83, Boulbonne, no. 80 (dated 1183, but probably from 1160s), fols. 154r ff., and Berman, "Growth," forthcoming.

[7] Cart. de Berdoues, nos. 92 (n.d. probably 1140s) and 125 (n.d.), and no. 173 (1152).

[8] Paris, B. N., Coll. Doat, vol. 70, Villelongue, nos. 2–7 (all dated 1149), and Constance H. Berman, "Dating Foundation Charters for Two Cistercian Abbeys in the Midi: Gimont and Villelongue," forthcoming.

[9] On Gerald of Salles, see Léon Chaigne, "L'implantation cistercienne en Poitou," Société d'Emulation de la Vendée, Bulletin 120 (1971):11–19; the edition of the Vita in Acta Sanctorum (Bollandists, 1861), Oct bx, cols. 249–54; and the edition by H. Martène and U. Durand, Veterum Scriptorum Amplissima Collectio (Paris: St. Maur, 1729), 6: 989–1014. For the links between Cadouin and southern French houses, see, Cartulaire de l'abbaye de Cadouin, précédé de Notes sur l'histoire économique du Périgord meridional à l'Epoque Féodale, ed. J.-M. Mabourguet (Cahors: Couselant, 1926), doc. 18, p. 19; Janauschek, Originum, 7–8, 108; Lacger, "Ardorel," cols. 1617 ff.; Pierre de Gorsse, L'abbaye cistercienne Sainte-Marie de Valmagne au diocèse de'Agde en Languedoc (Toulouse: Lion, 1933), 8 ff.

[10] See Toulouse, A. D. Haute-Garonne H205 Lespinasse, liasse 12; Mundy, Liberty, 16–17; Victor Fons, "Monastères," 116; Janauschek, Originum, 75; Cart. de Cadouin, 18–19, and Constance H. Berman, "Men's Houses, Women's Houses: the Relationship between the Sexes in Twelfth-century Southern France," forthcoming.

[11] Toulouse, B. Mun., MS 152, Grandselve, no. 6.

[12] Angleton, "Two Missions," 144, 171; and R. I. Moore, "St. Bernard's Mission in Languedoc," Bulletin of the Institute of Historical Research 47 (1974): 1–14.

poration in the 1140s, Fontfroide had gained an important reputation and considerable properties, including daughter-abbeys in Spain.[13] Later that abbey added to its patrimony by the incorporation of smaller priories.[14]

Other monasteries, although founded *de novo*, also profited from their absorption of nearby priories. Léoncel incorporated a priory at Pardieu to which its monks retired during the most unseasonable weather of winter.[15] Boulbonne's incorporation of the priory of Tramesaigues (which became Boulbonne's early-modern abbey site) became a *cause célèbre* and the subject of numerous arbitrated settlements because of resistance by the earlier mother-house, St. Michael of Cuxa, to the incorporation.[16] Cuxa's abbots continually attempted to regain control of Tramesaigues throughout the thirteenth century, but Boulbonne was generally successful in its legal proceedings against them and other Benedictines in the area.[17] In one such case Boulbonne was even accused of "fixing" the judges of arbitration because more than half were Cistercian.[18] Boulbonne also successfully annexed hermits in the forest of Boulbonne, as mentioned earlier, and a priory at Vajal which had been founded by reformers linked to Gerald of Salles.[19]

Such annexations of existing religious groups and their goods, either at the time of foundation or later, cannot fail to have given certain Cistercian houses an advantage over other new abbeys both in terms of land and personnel. Thus, houses like Grandselve and Fontfroide, two of the most successful abbeys of the south, may have prospered because they had an earlier "history" as independent religious houses before Cistercian incorporation. To count the growth of the order only from the Cistercian incorporation of such houses as is usually done, without taking into account such pre-Cistercian growth, tends to exaggerate the impact of Cîteaux as opposed to that of the larger movement of monastic reform.[20]

---

[13] E. Cauvet, *Etude historique sur Fonfroide* (Montpellier: Séguin, 1875), 173–86.
[14] Ibid., 226–50.
[15] *Cart. de Léoncel*, no. 53 (1194).
[16] In 1129 the abbot of St. Michael of Cuxa incorporated the independent priory of Tramesaigues (founded in the late tenth century); Paris, B. N., Coll. Doat, Boulbonne, vol. 83, no. 1 (962), etc.
[17] In 1209, St. Michael of Cuxa's abbot granted the priory of Tramesaigues to Boulbonne for the annual rent of twelve *solidi*; ibid., no. 138 (1209), and no. 139 (1209). It is probable that Tramesaigues was among property confiscated during the Albigensian Crusade and then restored to Boulbonne by the Cistercian bishop of Toulouse, Foulque, for ibid., no. 167 (1227) records the gift of Tramesaigues to Boulbonne by that bishop. This gift was confirmed by a bull of Gregory IX: ibid., vol. 84, fol. 16.
[18] Ibid., vol. 84, no. 170 (1230), fol. 28, in which the abbot of Saint-Anthony of Pamiers claims that too many Cistercian abbots were among the arbitrors.
[19] See Barrière-Flavy, *L'abbaye de Vajal dans l'ancien comté de Foix* (Toulouse: n.p., n.d.), passim.
[20] From very early on the order tended to simplify its own history, ignoring preceding groups and assigning "foundation" dates which were often dates of incorporation. Without this "prehistory" the expansion in the age of Bernard seems much more dramatic than it was in actuality, for many abbeys brought considerable endowment and membership with them. Fontfroide is certainly a good case in point, but it is not the only one; for further discussion of this point, see Berman, "Growth," forthcoming, and Jean-Baptiste Auberger, *L'unanimité cistercienne primitive: mythe ou réalité?* (Achel, Belgium: Editions Sine Parvulos, 1986).

In terms of property, the contribution of such priories to the monastic patrimony generally amounted to at least the core of one new grange.[21] Although such incorporations seem invariably to have been advantageous, an incorporation or site-change did sometimes cause donors to make special demands on the monks: for instance, that services be maintained at an earlier church or altar, or even that a group of monks or lay brothers remain at the old priory or hermitage site.[22] In terms of personnel, although both choir monks and lay brothers entered the order after being members of those earlier houses, it is rarely possible to assess the effect of such incorporation on the size of any community.[23]

Obviously, relationships with the neighboring laity were also crucial to the greater success of some houses over others. Local land-acquisition documents often reveal the relations between the new monks and the nearby community which provided the early donors, patrons, and eventual recruits to the order. Although later legend would often credit regional notables such as the Raymonds of Toulouse or the counts of Béarn with the foundation of Cistercian houses in southern France, initial donations of land more often came to the Cistercians from the men and women of a local knightly class which had arisen since Carolingian times.[24] Although religious

---

[21] Vajal and Tramesaigues were transformed by Boulbonne; Paris, B. N., Coll. Doat, vol. 83, Boulbonne, nos. 102 (1193), 104 (1195), and 105 (1195) and notes 16 and 17 above; the incorporation of Artigues by Berdoues, cited in n. 7 above; and Villelongue's incorporation of a hermitage which became its grange of Campagnes; see Paris, B. N., Coll. Doat, vol. 70, Villelongue, passim and Berman, "Dating." On the incorporation of la Pardieu by Léoncel, see n. 22 below.

[22] Donors were concerned that Pardieu continue to be used: *Cart. de Léoncel.* no. 53 (1194): "Hec autem domus, que Pars Dei dicitur, cum domo de Lionczel est eadem et unica abatia: unde conventus de Lionczel tenetur ibidem, scilicet, in Parte Dei, residentiam facere a festo sancti Andree usque ad Pascha, et per reliquum anni spatium debent ibi iiii vel sex de monachis in Dei servitio permanere." Similar concern was shown regarding the earlier church when Belleperche was moved to a new site at la Roque in the 1160s; Paris, B. N. Coll. Doat, vol. 91, Belleperche, nos. 1–4 (1164–66).

[23] Berdoues's incorporation of the hermitage of Artigues brought with it at least four members; *Cart. de Berdoues*, no. 173 (1152): "Et facto hoc dono receptus est Bertrandus de Marrencs major per monachum in domo Berdonarum. Et est sciendum quod quando predictus Bertrandus de Marrencs receptus fuit in domo Berdonarum per monachum tunc Sancius de Comeres et Johannes frater ejus et Ramundus Sancius de Cortade et Vitalis Destipuei et Vitalis de Maceriis donaverunt et concesserunt Arnaldo abbati Berdonarum et omnibus fratribus. . . . quidquid juris habebant in ecclesia Dartigas, et in ecclesia Sancti Saturnini, sive pro dono seu alio quolibet modo, ut fratres Berdonarum habeant et possideant, jure perpetuo totum quod ad ipsos aliquo modo in prenominatis ecclesiis pertinebat. Et hac donatione et concessione facta, Ramundus Sancius de Cortade est receptus a predicto Arnaldo abbate Berdonarum per monachum in domo Berdonarum. Et Sancius de las Comeres et Johannes frater ejus et Vitalis Destipuei et Vitalis de Maceriis, fuerunt recepti a predicto abbate per fratres et familiares in domo Berdonarum." Documents of conveyance to Artigues before incorporation amount to nine out of a total of ninety or ten percent of all conveyances for Artigues. See *Cart. de Berdoues*, nos. 173–266, especially nos. 189, 224, 229, 231, 232, and 246 (all undated).

[24] On the foundation of l'Escaledieu by the counts of Béarn, see Janauschek, *Originum*, 47; on Candeil, ibid., 132. The attribution of the Cistercian foundation of Candeil to a count Raymond of Toulouse seems to have resulted from the fact that Count Raymond VII was required to rebuild that church as part of the terms of the Peace of Paris of 1229; see *Histoire générale de Languedoc*, vol. 8, col. 886: "Abbatie Candelii CC marcas ad dicta monasteria construenda tum pro damnis eisdem." Efforts to associate Raymond Trencavel, viscount of

motives for the foundation of these new monastic houses are often cited in the charters, the pious acts of these men also legitimized their expanding power, as well as providing them an outlet for religious aspirations which the older, more aristocratic foundations may well have refused them. It was from this new class of knights, rather than from the older comital families, that most of the order's early recruits for choir monks came; from their tenants, peasants were recruited as lay brothers.[25] Indeed, most Cistercian foundations in southern France were instigated by men of these local knightly families, and almost always by laymen rather than clergy.[26]

Carcassonne, etc., more closely with the foundation of Valmagne seem to have been made by early thirteenth century, probably by Trencavel family members anxious about accusations of heresy rather than by the monks; at least one document in the Valmagne cartulary (redacted ca. 1210) was tampered with, perhaps in order to attribute the foundation of that house to him; see Montpellier, A.D., Herault, film, "Cart. de Valmagne," vol. 1, fol. 127r, charter dated 1138. Other sources suggest that the foundation of Valmagne must really be attributed to Raymond Trencavel's mother Cecilia, countess of Provence. On such female patrons, see Berman, "Growth," forthcoming. On the origins of the ruling class of the High Middle Ages in southern France, see Archibald R. Lewis, *The Development of Southern-French and Catalan Society* (Austin: University of Texas Press, 1965).

[25] Cistercian houses in southern France rarely drew either their leaders or members from outside, and they remained closely tied to the local communities in which they had been founded. This local emphasis may explain why the order was so successful in amassing the property which it did. In an area with little new land from which to create Cistercian estates, the order's ability to enter the land market knowledgeably (because administrators were local men) must have been very important. The earliest gifts made by such local knights and castellans were often crucial in determining the direction which monastic land acquisition took; that class also offered the monks its military protection. In contrast the most important men of the region, such as the Raymonds of Toulouse, tended to convey to the early Cistercians rights such as market privileges or freedom from tolls, rather than real estate; the untitled men of the knightly class were the ones who gave land to the Cistercians. For peasants, examples were Bernard of Bez, who with his son Bernard gave the *pagesia* of the *mansus* of Bez to Radulfus abbot and Bonnecombe, where he would be received as *conversus* if his wife allowed: *Cart. de Bonnecombe*, no. 264 H (1183); and Peter of Caucill who gave half of a *pagesia* of one *mansus* and the *appendaria* of Caucill and one quarter of the *pagesia* of the other *mansus*. He was received as *frater* and *conversus* and his sons confirmed the concession, ibid., no. 268 C (1187). Deodatus Jovus with his wife and daughter gave *pagesia* at Carabaza and Landola to Bonnecombe. He was to be received as a member if he made his entrance to the house within the next three years, although his status was never revealed: ibid., 273 C (1193). Knights made their gifts for entrance into Bonnecombe in acts such as ibid., no. 274 D (1194) in which Ugo Bonefons owner of *fevum* and *allodium*, etc., entered, or possibly ibid., no. 269 B (1188). For the sons of knights becoming monks, see *Cart. de Silvanès*, no. 199 (1158), becoming *conversi*, ibid., no. 179 (1158), an indeterminate status, ibid., no. 183 (1158); 160 (n. d.); and sons were received as monks while parents as *conversi*, ibid., no. 202 (1158) and Rodez, Soc. des lettres, Bonnecombe, "Cart. d'Iz et Bougaunes," no. 112 (1194), in which knights entered.

[26] Several examples from the Bonnecombe cartularies show wealthier members of the surrounding community becoming important administrators in the abbey. Ugo of Salvinniac, for example, became a *conversus* at Bonnecombe and acted as witness in a number of charters or received gifts for the abbey; see *Cart. de Bonnecombe*, no. 167 (1181), 173 (1186), 264 H (1183), 267 C (1186), and 269 D (1188), and similarly William of Montpellier, as monk acting as witness, ibid., no. 251 C and D (1166).

Knights holding castles in the vicinity of Bonnecombe and other abbeys were donors and often entered the order; for example, such men as Gago of Peyrebrune: ibid., no. 162 (1174); Ymbert of Castelpers, ibid., no. 163 (1174); Montarsin of Calmont, ibid., no. 166 (1180); Raterius of Solis, ibid., no. 178 (1206); Ugo of la Roca, ibid., no. 179 (1206); Berengar of Auriac, ibid., no. 182 (1211); Peter of Turre, ibid., no. 188 (1215); Hector of Mirabel, ibid., no. 189 (1215); Deodat of Caylus, ibid., no. 190 (1216); Raymond of Castronovo, ibid., no.

The foundation of the Cistercian abbey of Bonnevaux in Provence in 1119 by the archbishop of Vienne, Guy de Bourgogne, just before he became Pope Calixtus II, is the exception.[27]

Around the time of foundation many property conveyances to the Cistercians were donations, but the order's consolidation of rights at virtually every grange was continued by careful land acquisitions aimed at improving holdings already in hand, and purchases represented the major means by which this land was acquired.[28] Such purchases, however, were often disguised as "donations" for which the monks gave a reciprocal gift in cash out of love (*in caritate* or *de caritate*) to the "donors."[29] As Table 2 shows, of all conveyances to the order (including confirmations of earlier grants) more than half were onerous in some way; in such contracts, the monks

---

206 (1229); and Arcambald of Panat, ibid., no. 214 (1240). The last was from the family of Panat which had been responsible for the foundation of Bonnecombe and also for that abbey's endowment with pasture rights; some family members became monks there: *Cart. de Bonnecombe*, no. 251 (about 1166), for the donations to Candeil and its abbot Gausbert for the founding of Bonnecombe by local knights.

At Silvanès, founders promised specifically to protect the vineyards of the monks at Laurs, *Cart. de Silvanès*, no. 243 (1166): "Et cognoscimus esse juris vestri ut de riparia Zatgairenca ligna sufficienter accipiatis ad construendam et reparandam vineam vestram de Lauro, et ideo debemus eam custodire et defendere ab incisione et exterminatione." They promised to defend them, ibid., no. 210 (1159) and granted them tithes upon departure for a pilgrimage to Jerusalem: ibid., no. 189 (1159).

[27] J.-M. Canivez, "Bonnevaux," *Dictionnaire d'histoire et de géographie ecclésiastique* 9 (1937): cols. 1074–76.

[28] Charters in the surviving archives for this period are sometimes loose parchments, copies preserved in codices of the twelfth or thirteenth centuries, or summaries included in later inventories; see Mireille Castaing-Sicard, "Donations toulousaines du Xe au XIIIe siècles," *Annales du Midi* 70 (1958): 27–64; and idem, *Les contrats dans le très ancien droit toulousain (X^e-XIIe siècle)*, (Toulouse: Espic, 1959), 20 ff.

Gifts were made to Gausbert abbot of Candeil for the foundation of Bonnecombe, *Cart. de Bonnecombe*, no. 251 B (c. 1166), and by Bernard Count of Astarac to Morimund to construct Berdoues: "Sciendum est quod Bernardus comes Astaracensis et Sancius dictus ejus filius . . . donaverunt, concesserunt et absolverunt Deo et beate Marie Morimundi et Wualtherio abbati et conventui ejusdem loci presenti et futuro terram de Berdonis et ecclesiam ejusque casale cum omnibus ex integro pertinentiis eorum quas habebant vel habere debebant per se vel per aliam personam cultas et incultas ut ibi construeretur abbatia cisterciensis ordinis": *Cart. de Berdoues*, no. 92 (dated 1134, this document should probably be placed after 1142).

[29] For example "M.C.LXXX.VIII. Ego Ugo de Seveirac, per me . . . , dono et concedo Deo et Beate Marie et monasterio Bonecumbe et tibi Bertrando, abbati ejusdem loci, atque vobis aliis fratribus presentibus et futuris, totum hoc quod habeo vel habere debeo in mansis Frigidarum Mansionum et in Carabaza et in Caucill et in Becio et si quid aliud habeo ab ecclesia de Tremoillas usque ad vestram abbaciam, videlicet sicut via descendit ab ecclesia predicta et transit per mansum, qui Ventago dicitur, usque ad Seor: quicquid, inquam, infra terminos istos continetur versus domum vestram, si quid, ut dixi, habeo vel habere debeo in toto territorio isto quod vobis hic dico et nomino, dono et desamparo vobis in perpetuum. Preterea dono vobis meas herbas atque herbatges et nemora mea ad usus vestros et juravi super sancta Evangelia tacta quod nichil in predicto dono queram vel querere cuilibet faciam et ero vobis jure defensor de hoc ac dono vobis fidejussores pro garencia Ugonem, comitem Ruthene, et Bernardum Gallardum. Vos autem dedistis michi in caritate C L solidos": *Cart. de Bonnecombe*, no. 269 B (1188), and see G. Chevrier, "Remarques sur la distinction de l'acte à titre onéreux et de l'acte à titre gratuit d'après les chartes du Rouergue au XIIe siècle," *Annales de la faculté de droit de l'Université d'Aix-Marseille* 43 (1950): 67–79. In Table 2, promise of admission into an abbey has been treated as potentially onerous, whereas burial had not been. The disguised sale is discussed by Castaing-Sicard, "Donations," 27–34.

paid cash in return for land, made a payment in produce, promised an annual rent in money or kind, or granted a potentially burdensome benefit such as the right to enter the abbey as monk or *conversus*. Occasionally, monastic purchases included the redemption of land under mortgage, and a number of holdings came into Cistercian hands because they had previously been mortgaged by the donor or vendor, although the monks were not normally the original lenders in such transactions.[30] Conveyances of land under mortgage were sometimes made without mention of that mortgage; in fact only residual claims to the land or rights to its redemption were actually acquired. This causes some confusion about the relative value and price of land conveyed to the monks by various individuals.

With the majority of their acquisitions made by purchase on the land market and only a minority of them in free alms, the Cistercians in southern France were forced to expend large sums on land acquisition in order to create the large consolidated granges on which their direct cultivation could be instituted, as indicated in table 2. Such land acquisition through purchase was probably made easier for Cistercians in southern France than elsewhere because the economy of the Midi was extremely monetized.[31] There was a considerable demand by the families who ruled the region for cash, reflected in their already mortgaged lands. Cash was needed not only for consumption, but for investment in town and industrial growth, and for military expeditions. Cash was the means of payment for all types of transactions: from payment of rents and services by tenant farmers to the transfer of estates between members of the princely families.[32] This use of cash is reflected in the charters, which rarely include property exchanges (which appear to have been almost entirely replaced in the twelfth-century Midi by multiple transactions in cash).[33] Cash values are mentioned even in those infrequent charters where payment was made by a cash substitute such as livestock.[34] Generally, however, cash values meant money payment. Indeed, a number of charters actually describe the counting out of *deniers*

---

[30] For example, Rodez, Soc. des lettres, "Cart. d'Iz et Bougaunes," nos. 72 (1183), 121 (1194), and 158 (1199); in the last certain lands ceded to Bonnecombe by both the knights of Panat and the counts of Rodez are revealed to have been under mortgage to those knights as surety for a loan of one thousand *solidi* which they had made to the counts of Rodez. There is similar indebtedness found in the Berdoues cartulary for the Montesquiou family in Gascony, but they borrowed from the Cistercians; see Berman, "Mortgage," 250-58. On the use of the mortgage contract in the Midi in general, see Castaing-Sicard, *Contrats*, 295-371.

[31] Ibid., 229-39.

[32] Use of cash, contracts and even mortgages, rather than "feudal oaths" as means of sealing relationships between individuals at the top of society was very prevalent in the south. I am grateful to Prof. F. Cheyette for bringing some of these relationships to my attention.

[33] Compare tables for southern France in earlier centuries in David Herlihy, "Church Property on the European Continent, 701-1200," *Speculum* 36 (1961): 81-105.

[34] For example, 35 goats worth 60 *solidi* in *Cart. de Gimont* I, no. 73 (1176). Rates of exchange of different coinages of their values are mentioned in *Cart. de Silvanès*, no. 167 (n.d.) which mentions payment of two *librae* at the rate of 95 shillings of Melgueil per pound of silver; or in Rodez, Soc. des lettres, "Cart. d'Iz et Bougaunes," no. 50 (1182), where fifty *solidi* of Rodez are described as worth one mark of silver.

in payment; there is no question of a non-monetary economy dealing in a money of account.[35]

As a whole, even though individual transactions were often small, the amounts expended by the order in its land acquisitions were considerable; (see Table 2). Individual payments ranged from the two or three *solidi* granted by Grandselve to a tenant-cultivator giving up his heritable rights, to the sixty *librae* (1200 *solidi*) paid by Bonnecombe to a local woman for rights over an entire nearby village, or the 4000 *solidi* paid by Valmagne for rights at Vairac.[36] The average price paid by any particular abbey for a single land purchase ranged from about twenty to well over two hundred *solidi*, although comparison is complicated by the fact that the values of the local currencies differed slightly.[37]

The charters reveal very little about the source of the cash which the order used in payments for land.[38] In the earliest years, cash may have come to the monks through money gifts made by local patrons,[39] and in

---

[35] Payment in *denarii* of the 350 *solidi* owed to Peter Ymbert for the *mansus* of Puy d'Oneth, ibid., no. 153 (1201), or see *Cart. de Bonnecombe*, no. 286A (1205).

[36] Purchases could be extremely large or extremely small. Bonnecombe's purchase of the grange of La Serre from Bonneval in the 1220s involved payment of 12,000 *solidi*; see *Cart. de Bonneval*, nos. 140–42bis (1225) and 151 (1232). On the smaller end of the scale, see Paris, B. N. Latin MS 11008 Grandselve, fols. 155r–56r (no. 74) (1165), in which four *solidi* of Morlaas were paid for rights in the territory of Gaugiac and for part of the *casal* of Boseville, or ibid., no. 75 (1165), in which two *solidi* were paid for rights over the territory of Gaugiac and part of the *casal* there. The latter donor, however, was also given confraternity rights. Similarly, Terrenus presbyter gave land at Pradela to Grandselve for two *solidi* of Cahors: Paris, B. N., Latin MS 11011, Grandselve, no. 4 (1147), fol. 2v; a tenant there received five *solidi* of Cahors and Bernard Terz and his family received four *solidi* of Cahors, loc. cit.; Gerald Espaniol gave Grandselve land at Lassela adjoining land which the monks owned, for two *solidi* of Morlaas, ibid., no. 36 (1160), fol. 24v. On the other end of the scale is *Cart. de Bonnecombe*, no. 299B (1225): "M.CC.XX.V., VI. nonas maii. Ego Aicelina, uxor Raimundi Talairan principaliter et ego Raimundus, maritus ejus secundario et nos Nobila, Bartolomeus, Ugo Petri et Willerma, eorum infantes, cognoscentes XIIII annorum nos esse majores, nos omnes . . . vendimus . . . domui Bonecumbe, fratribus seu monachis presentibus . . . et vobis fratribus Raimundo Carbonelli, cellarario et Willermo Foramont, pro ipsa domo, . . . videlicet: totum mansum de La Calm pro feudo et alodio et medietatem tocius mansi dal Dairese et quartam partem mansi de Teuleiras et totum mansum de La Brugaireta pro feudo et alodio et apud Sanctum Ylarium duos ortos et duas casals et in Restapauc duos mansos pro feudo et medietatem trium parcium tocius mansi del Pojet et feudum mansi de La Malaira et quicquid juris habemus, habere possumus vel debemus in dictis honoribus a fluvio de Biaur usque ad Solmeih cum omnibus juribus, pertinenciis et apendiciis suis. . . . Propter hanc autem venditionem . . . , vos dicti fratres dedistis nobis LX libras Ruthenensis monete bone et percurribilis." Witnessed by the bishop and canons of Rodez.

On Vairac, see Montpellier, A. D. Hérault, film, "Cart. de Valmagne," vol. I, fols. 100–35, and vol II. fols. 131–46.

[37] On values of various currencies in the region, see Mireille Castaing-Sicard, *Monnaies féodales et circulation monétaire en Languedoc (Xe-XIIIe siècles)* (Toulouse: Assoc. Marc Bloch, 1961), esp. 31–35.

[38] If amounts given in wills such as those cited in the next notes were actually paid (probably some of the bequests promised by the last Raymonds of Toulouse were not), such gifts represent one source of the funds for land purchase. Unfortunately, documents do not mention where funds came from.

[39] The Silvanès chronicle does list donors who made large grants or who put up various buildings, see *Cart. de Silvanès*, no. 470, p. 386, but the list includes so many improbable names and should probably be discounted altogether. Wills found in the *Histoire générale de*

particular, by way of testamentary bequests.[40] There is, however, no evidence that the bequests in surviving wills were always paid and it is unlikely that the amounts were sufficient to meet the enormous expenses of Cistercian expansion as carried out in the period up to ca. 1250. The documents do not suggest that the successful Cistercian houses of the region resorted to borrowing in order to meet expansion costs; indeed, smaller houses, like Locdieu, which were forced by circumstances to borrow, can be shown to have done so to their grief.[41] Generally speaking, borrowing against the mortgage would not have aided land acquisition in the long term, because of the severe terms of that contract in the Middle Ages.[42] Moreover, unlike Benedictine houses in the region, such as Moissac and Conques, which became major stages on the pilgrimage routes to Spain and actively promoted the cult of their relics, the Cistercians in the Midi seem to have sought no pilgrim trade.[43] Efforts to advertise the holiness of at least one

---

Languedoc and elsewhere, include some substantial gifts to Grandselve and Valmagne among others from counts of Toulouse and lords of Montpellier. See the following notes.

[40] The wills of the various Williams of Montpellier include bequests to houses of the Cistercian order; see *Liber Instrumentorum Memorialem. Cartulaire des Guillems de Montpellier* (Montpellier: Jean Martel Aîné, 1884–86), no. 96 (1172): "In nomine Domini Jhesu Christi, et gloriose virginis Marie. Anno ab incarnatione Dominica M.C.LXX.II., mense septembris, in festivitate sancti Michaelis. Ego Guillelmus, dominus Montispessulani, filius quondam Sibilie, mea bona memoria sic testamentum meum facio, et ultimam voluntatem meam super rebus meis dispono. In primis dimitto corpus meum ad sepeliendum in monasterio Sancte Marie Grandissilve, et eidem monasterio relinquo pro monacho Raimundum, filium meum, et ei pro eo, ipsi eidem monasterio Grandissilve, M. solidos Melgorienses relinquo, quibus Raimundum, filium meum, contentum esse volo. . . . Dimitto operis ecclesie Vallismagne D. solidos Melgorienses, et mando quod compleatur illud relictum, quod demisi ad opus dormitorii ejusdem ecclesie monasterii—Dimitto monasterio Francarumvallium D. solidos Melgoriensis."

Also no. 97 (1177): "Anno ab incarnatione Domini M.C.LXX.VII., mense februarii. Ego Guido Guerregiatus, in mea bona existens memoria, sic ultimam voluntatem meam facio, et rebus meis dispono. Reddo et dono me ipsum Deo, et Beate Marie Vallismagne, in ordine Cisterciensi. In primis dono et laudo molendinos de Paollano, et terram de Vallautre, et terram de Cocone, vivam aut moriar, monasterio Vallismagne, in sempiternum"; and no. 99 (1202): "In nomine Domini. Anno incarnationis ejusdem M.CC.II., pridie nonas novembris. Ego Guillelmus, Dei gratia Montispessulani Dominus, filius quondam Mathildis ducisse, in mea bona memoria, et in ultima voluntate mea, sic dispono et ordino testamentum et ultimam voluntatem meam. . . . Et volo et jubeo heredi meo, domino Montispessulani, ut faciat corpus meum defferri et sepeliri in cimiterio Grandissilve, cui monasterio Grandissilve dimitto C libras inter opus et mensam domitorium." The Peace of Paris of 1229 with bequests by Raymond of Toulouse is cited in n. 2 above; whether such bequests were actually paid is more difficult to determine.

[41] Bonnecombe borrowed money from Montarsinus of Calmont in 1178 against a mortgage; the mortgaged land was returned to the abbey without payment in the lender's will and that transfer was confirmed by his heirs; see *Cart. de Bonnecombe*, no. 166 (1180), and Rodez, Soc. des Lettres, Bonnecombe, "Cart. de Magrin," nos. 16 (1178), 28 (1182), and 29 (1182). This is the only case in which the monks of that house appear to have borrowed. Locdieu's difficulties with Bonneval ultimately seem to stem from borrowing; see chapter 5, pp. 112–13.

[42] The terms of the medieval mortgage required that the fruits of the land be transferred to the lender until the mortgage was repaid; this made it particularly difficult to redeem; see Robert Génestal, *Le rôle des monastères comme établissements de crédit: étudié en Normandie du XIe à la fin du XIIIe siècle.* (Paris: Arthur Rousseau, 1901) and Castaing-Sicard, *Contrats*, 295–371.

[43] Economic results of the cult of relics which fostered the pilgrim trade are discussed by Patrick Geary, *Furta Sacra: Theft of Relics in the Central Middle Ages* (Princeton: Princeton University Press, 1978), 24–25; Geary gives no references to Cistercian relic cults or pilgrimage to Cistercian churches.

foundation, however, were obviously part of an attempt to attract donations.[44]

Since the order lacked such sources of cash as borrowing, or pilgrim trade, and received few enormous bequests, it must be concluded that from very early in its history the cash for land purchases was internally-generated. The capital needed for expansion must have been provided by the savings from monastic frugality and by profits from the order's earliest agriculture and pastoralism. It is obvious that once grange farming had been introduced, cash to purchase land could have come from selling surplus cereals, but the problem is how the first granges would have been paid for before there were profits from cereal production. It is probable that in most cases, production of surplus grain for sale would only have begun after the order had made considerable land acquisitions. This means that land must have been purchased at the outset with the profits from the sale of animals and animal products, and possibly from the profits of milling flour.[45]

It is difficult to know how soon revenues came to Cistercians from milling, but those from animal husbandry were probably available from almost immediately after a new Cistercian house was founded. Some of the cheese, butter, and milk which the monks produced would have been consumed by them, as would some wool and leather for clothing, harnesses, and other equipment, but Cistercian dietary restrictions and ascetic practices meant that pastoralism could have quickly provided marketable meat that the monks did not consume, and horses that they did not ride, as well as surpluses of other animal products. If this conclusion (based on logical argument rather than on documentary evidence) is true, it is likely that Cistercian success in pastoralism was almost everywhere a necessary condition for later success in grange agriculture. The most successful houses of the order in southern France were probably first and foremost successful pastoralists, because Cistercian houses without sufficient pastoral resources probably never had the cash to create the huge granges which would later provide grain to regional markets. This conclusion is confirmed by the evidence of the charters inasmuch as they show that all the most successful Cistercian houses of the region eventually amassed considerable pasture rights which were carefully maintained. Once the monks had cash available for land purchases, the slow and painstaking process of land consolidation and reorganization began. Through this rationalization of land-holdings many Cistercian houses of monks in southern France successfully developed a cluster of large granges which supported them in later centuries. In the process, they transformed the countryside and the rural economy which had preceded them.

In their reorganization of agricultural land the monks had not only the practical advantages of privileged status: exemptions from market tolls,

---

[44] This must have been a major reason for the composition of the Silvanès chronicle which attempts to advertise the holiness of the abbey's founder, Pons de Léras.

[45] On mills as a souce of early revenue, see Georges Duby, *Bernard de Clairvaux: L'art cistercien* (Paris: Arts et Métiers, 1976), 107.

tithe privileges, and other features discussed in chapter 4, but an ideology which promoted savings, efficiency, careful planning, and hard work. Abstinence in dietary matters has already been mentioned, but the Cistercian desire for simplicity also prevented lavish expenditures on building, decoration, church plate, and liturgical display. Their ideals also kept the monks working days and all of their labor to prayer.[46] Manual labor was very important to the order, even if in southern France it did not extend to clearance and reclamation. For Cistercians to work with one's hands was to pray, and, at least in the early years, even the choir monks must have worked alongside the *conversi* in the fields. Not only did this mean that fewer "unproductive" mouths had to be filled, but it also meant that the monks themselves became closely involved in all aspects of land management and acquisition.

Their longevity and written memory also gave the monks a tremendous advantage in land acquisition over their lay neighbors who resorted less often to the written charter or cartulary; the monks were able to wait for years, if necessary, to acquire a particularly valuable piece of land, or await arbitration in land disputes.[47] Thus, associated with the order's use of written documents were legal skills and the wealth necessary to carry on protracted disputes which may well have bankrupted their opponents. The cartularies are full of accounts of arbitrated legal settlement (or supposed settlement) of disputes over land or pasture rights, often in cases initiated by the monks.[48] In a few instances they even mention that the monks brought

---

[46] Early Cistercian statutes, for instance, forbade the use of gold crosses and plate for the altar, or the construction of ostentatious churches, and prescribed simple dress; see *Statuta*, ed. Canivez, vol. I, 1134, nos. 4, 10, 20, etc. Judging from surviving church buildings of the twelfth century, such as Silvanès, Sénanque, and Flaran, these aims were maintained at least in the earliest period: Marie-Anselme Dimier, *Art Cistercien*, trans. Paul Veyriras, 2nd ed. (Paris: Zodiaque, 1974), passim.

[47] Monastic cartularies are among the earliest surviving volumes of documents for this region. Professor Paul Ourliac (personal communication) has suggested that many were compiled as a result of uncertainty in tenure and other insecurities of the period of the Albigensian Wars. For a discussion of them in this region, see Castaing-Sicard, *Contrats*, 23 ff.

[48] The Boulbonne dispute with Cuxa over Tramesaigues has already been mentioned. Recorded disputes, as mentioned in surviving *convenientiae* which record their settlements for the 1150s and 1160s, are the following:

date	abbey	subject	reference
1154	Silvanès—Hospitallers	tithes	*Cart. de Silvanès*, nos. 170, 171, 174 (1165)
1158	Gimont—Grandselve	pasture	*Cart. de Gimont* III, nos. 50 and 66
1161	Silvanès—Valmagne	grange	*Cart. de Silvanès*, no. 303.
1165	Grandselve—Moissac	land	Paris, B.N. Latin MS 11011, Grandselve, no. 87.
1166	Bonnefont—Templars	land	*Rec. de Bonnefont*, no. 80.
1166	Boulbonne—Combelongue	pasture	Paris, B.N., Coll. Doat, vol. 83, Boulbonne, no. 40.
1168	Grandselve—la Capelle	land	Paris, B.N. Latin MS 11008, Grandselve, no. 94.

forward the charters at issue in such disputes, or in cases where they wanted to encourage heirs of the original donors to confirm the gifts of their relatives.[49] In addition to their legal skills, abbeys of the order also had a moral advantage over those from whom they desired to gain land rights. Bequests were sometimes made at the burial of a beloved parent, or in the last moments of a donor's life, when the monks hurried to his death-bed or brought him during his final illness to the monastic infirmary.[50] When violence or aggression was offered the monks or resistance to their expansion was made by knights or peasants, such responses often ultimately led to monastic victory. Murderers or thieves, in fear of eternal punishment, often made amends in the last moments of their lives for wrongs done to the monks—often conceding the very land which the monks had originally sought.[51] Finally, it did no harm to the order's land acquisition in the late twelfth and early thirteenth centuries that Cistercians were closely associated with papal preaching missions and Crusades against heresy in the region.[52]

Such advantages inherent in both the Cistercian way of life and in the order's position within Christian society were important in monastic land acquisition, but much more significant were the planning for and management of land acquisition exhibited in the process of charter and cartulary redaction itself. The charters of land acquisition were important documents which helped the monks retain holdings once gained. When sorted and eventually redacted into books of charters or cartularies, they also enabled the monks to evaluate their position in the land acquisition process, allowing them to assess what had been purchased, what remained to be purchased,

---

[49] See, for example, *Cart. de Berdoues*, no. 352 (1242): "Preterea notandum est quod predictus Petrus de Orbezano concessit firmiter et firmavit fideliter in conspectu et in presentia domini Centulli comitis Astaraci illud generale donum de pasturis et illam espleitam pascuorum quam in communi et solemni astaracensi curia Gillelmus Gassie de Orbezano avus suus donavit Deo beate Marie et habitatoribus domus Berdonarum in presentia domini Gillelmi auxitani archiepiscopi et Bernardi comitis Astaracensis, militibus, in communi carta continetur" and ibid., no. 371 (1245), which describes the division of the charters *per alphabetum*.

[50] Burial was promised to Sanche Lub of Faisan who made a gift "in infirmitate illa de qua mortuus est"; he was promised reception and burial in the abbey as a monk and honorable burial as a member of the community at Berdoues, ibid., no. 563 (n.d.). Donations for reasons of health or in face of death also include the grant of the church of St. Stephen of Durfort to Berdoues, ibid., no. 459 (1211), cited in note 78 below, or ibid., no. 761 (n.d.). A gift was made for the soul of Arnold of La Serra (who had died and was buried at Gimont) by his sisters Narbona and Gasena and by the children of Narbonna: *Cart. de Gimont*, II, no. 189 (n.d.). Odo of Lomagne, viscount, gave gifts and requested burial at Grandselve: Paris, B.N., Coll. Doat, vol. 76, Grandselve, no. 124 (1237); Ugo of Durfort and his three daughters made a gift to Villelongue for the burial of his wife and their mother: Paris, B.N., Coll. Doat, vol. 70, Villelongue, fol. 149r (1244).

[51] *Cart. de Berdoues*, no. 655 (n.d.): "Iterum sciendum est quod Lauzig de Stelan, qui dicitur Mula, cepit oves fratrum de Berdonis malicia et superbia sua, quas redemerunt fratres de Berdonis tres sol. Tandem ille ad mortem suam recognescens malum et peccatum suum ordinavit et concessit, firmavit illos tres sol. in decima terre de Samaran"; ibid., no. 470 (1154) records the concession of lands by villagers after one of their number had killed a monastic servant. See below, pp. 58–59.

[52] Some benefits are mentioned in chapter 6, p. 124.

how much they had spent, how much would need to be spent for similar acquisitions, and which individuals still had claims to their land.[53] The cartularies clearly affirm the managerial prowess of the new order, showing how the monks methodically sought holdings which would enhance land already in their possession, and confirmations of their acquisitions from all possible claimants. By adopting careful acquisition procedures, it was possible for Cistercians to amass large, compacted holdings which were transformed into granges on which they were the only owners, despite the lack of wilderness land in southern France. This was accomplished by the monks' careful purchases of all rights to each parcel of land from all existing owners. It required skill in awaiting the proper moment to solicit donations or purchases, and in bringing together the sums necessary for important purchases, but monastic houses could afford to be patient.

The purchases and acquisitions which most enhanced monastic property were of two major types which reflect the types of fragmentation discussed in chapter 2: those which added new areas of land adjacent to lands already owned by the monks—in effect, reversing the fragmentation of actual holdings—and those which reduced the number of claimants to the produce of individual land holdings already in monastic possession. Although many of the charters of land conveyance to the monks can be shown to fall into one of these two types of consolidation patterns, there are also charters in which both types of consolidation occurred at once. The rationalization and planning involved were obviously, however, much more varied and responsive to local conditions than that described here, and the time over which such acquisition occurred might vary considerably.[54]

The first type of land rationalization, the effort to acquire coterminous holdings, or those intermingled among Cistercian fields and farms, can be called "compacting." This process occurred at a variety of levels of ownership and on different scales. Occasionally, it involved the purchase of new villages adjoining those already owned, but more often it occurred on a smaller scale: for example, by the addition of new farms, new fields, or even new strips of land to those already held by the monks. Once holdings in a single area were put together, acquisitions expanded into additional villages or estates by way of the single detached farms which filled the interstices between the villages, and then by acquisition of single holdings

---

[53] The cartulary of Gimont is typical, having six major sections concerning respectively: 1) the abbey and its home farm, 2) the grange of Laurs, 3) the grange of Hour, 4) the grange of Franqueville, 5) the grange of Saint-Soulen, and 6) the grange of Aiguebelle. The whole was probably copied no earlier than ca. 1220 with a few additions to the charters for Laur in the years up to 1233. Bonnecombe had separate cartularies for each of its granges and it is possible that copying was done directly into the cartulary books at the granges by Bonnecombe's monks and grangers, since acts in the cartularies have a variety of signatures.

[54] Acquisitions could occur as slowly as those for Berdoues's grange of Bedored, see *Cart. de Berdoues*, nos. 700–45, which extend over a century, or as quickly as the more that 350 acts in the Silvanès cartulary which concern property acquired in less than 40 years.

in the new village or parish.⁵⁵ Only gradually did an abbey acquire significant holdings in the more peripheral areas, tending instead to concentrate its cash and attention first on the consolidation of rights in the more compacted center.⁵⁶

Most evidence of land compacting comes from areas where land conveyed to the order was described in contracts giving the perambulations of a holding: either the geographical boundaries of one or more of its four sides (such holdings were generally rectangular), or lists of the contiguous owners of land. Such descriptions of holdings by either part or all of their perambulations are most often found among contracts from the more Romanized areas of the study, particularly in documents conveying land to the Cistercians in Languedoc. Some examples can be cited from documents for Valmagne's acquisitions for its grange of Mercurières near Montpellier. There Pons Bertold conveyed to the monks a field at Pardiniac, which adjoined a field belonging to Raimonde of Milsan on one side and a field belonging to William Marcel on the other.⁵⁷ Similarly, two pieces of land which adjoined fields belonging to Raymond Bedaz and to William Herbert were donated to the monks by Peter Abole and his sons.⁵⁸ Aimeric of St. Paragoire conveyed land which adjoined the grange of Mercurières.⁵⁹ Gerald of la Paille of Mèze with his wife and two daughters gave land at Mercurières adjoining the field belonging to Pons Martin on one side and the road to Genciac on the other,⁶⁰ and so on. In this process of compacting, the most valuable holdings were obviously those which adjoined lands already held by the monks on one or more sides. Thus, very important to this compacting at Mercurières was the conveyance by Peter Durand of Bézins in which he transferred to the monks land at Mercurières "which was adjoined on all sides by territory already belonging to Valmagne except on the side which adjoined the road from Bézins to Creis."⁶¹ Such compacting is similarly explicit in documents such as that recording the conveyance to Candeil of the *bordaria* of Jourdain which adjoined the *mansus*

---

⁵⁵ Holdings at the *villa* of Oneth were acquired because they adjoined the grange of Iz already belonging to Bonnecombe; Rodez, A. D. Aveyron, 2H Bonnecombe 39-1, no. 14bis (n.d.).

⁵⁶ Similarly, the *appendaria* of Parlant, or the isolated *mansi* of Cairo, Cairaguet, and Rover, were later additions by Bonnecombe to its properties in and adjoining the *villa* of Magrin: Rodez, Soc. des lettres, MS Bonnecombe, "Cart. de Magrin," nos. 80 (1192), 64 (1195), and 15 (1178), 89 (1202), 56 (1204), 54 (1205).

⁵⁷ Montpellier, A. D. Hérault, (film), "Cart. de Valmagne," vol. I, fols. 41r–42v, three acts dated 1168, and also no. 6 (1155–72) 41r.

⁵⁸ Ibid., no. 154 (1155–72) fol. 43v.

⁵⁹ Ibid., no. 8 (1162) fol. 41v.

⁶⁰ Ibid., no. 10 (1163) fol. 42r.

⁶¹ Ibid., no. 18 (n.d.) fol. 44r; similarly see for Fontfroide, Carcassonne, A.D. Aude, H 211 "Inv. de Fonfroide," fol. 38r, in which sides of conveyances are described by the prevailing winds from those directions, or identified by giving the names of the owners of honors, or roads on each adjoining side.

CONSOLIDATION OF RIGHTS   45

of Jourdain which had already been acquired by the monks.[62] Such a *bordaria*, generally a smaller holding than a *mansus*, was, like the *appendaria* in the Rouergue, an addition to already old tenures made before Cistercian acquisitions in the area.[63]

Probably the best example of such compacting activity comes from Grandselve's documents. They show the process of such rationalized acquisition particularly well for the areas of relatively new settlement along the Garonne and Gimone Rivers north of Toulouse.[64] For example, in 1147, at the site near the Garonne of what would later become Grandselve's grange of Bagnols, Arnold Helye and his son conceded their land with its tithes and first fruits to the monks in a conveyance which was confirmed by the religious men of St. Sernin in Toulouse.[65] This holding is described as adjoining St. Sernin's land on the east and the old channel of the Garonne on the west, although no northern or southern owners or boundaries are mentioned. In a slightly later contract, Peter of Bagnols gave land to the monks which was located directly north of the first holding; that is, it adjoined the land of St. Sernin in the east, the old Garonne channel on

---

[62] Paris, B. N., Coll. Doat, vol. 114, Candeil, no. 29 (1168) fol. 58 ff. and no. 40 (1169), fols. 78 ff, and no. 18 (1169), fol. 29 ff.

[63] Bordaria is defined in Niermeyer, *Lexicon*, p. 101, only as a bordar's holding.

[64] Consolidation along the Gimone River is found in Paris, B. N., MS 11009, Grandselve, passim and Paris, A. N., L 1009 bis, Grandselve, passim. Almost all these charters concern land described simply as between "the Gimone, Serrampione, and Auzi Rivers," which was conveyed to the monks of Grandselve. See also *Cart. de Gimont*, III, no. 50 (1158) and ibid., no. 66 (1158–74), the controversy over grange sites between Gimont and Grandselve: "Siquidem abbas Gemundi asseruit his rationibus abbatem Grandissilve non debere hedificare grangiam in prefato loco, primo cum ibidem domus Gemundi habuerit de Gallardo de Sirac locum et sedem ad construendam grangiam prius, transactis amplius quam XII annis; secundo quod quidam Roaldus juxta ibidem acquisiuit locum satis aptum ad grangiam hedificandam, ubi et habitationem cum auxilio et consilio fratrum Gemundi construxit et quedam seminavit ibi in loco qui vocatur Silva Arriana et hunc locum postea fratribus Gemundi dedit; tercio quia infra terminum grangie que vocatur de Francavilla et grangia que vocatur del Forc et ipsius abbatie esset prefata grangia de Tarrida." The second reason, that the monks aided a certain Roald to build and sow at that place, may possibly refer to clearance undertaken by a peasant-cultivator, under the aegis of the monks of Gimont, of land which afterward was transferred to the monks.

[65] Paris, B. N. Latin MS 11008, Grandselve, no. 1 (1147), fol. 1v.: "m.c.xi.vii. . . . Ego Arnaldus Elie et ego Odo filius eius per nos et per omnes nostros successores sine omni retentione donamus et concedimus in perpetuum pro bene et fide domino Deo et Beate Marie Grandissilve et vobis fratribus eiusdem loci presentibus et futuris pro salute animarum nostrarum et omnium parentum nostrorum cum consilio et voluntate Ugonis Sancti Saturnini abbatis et canonicorum eiusdem ecclesie, condaminam nostram de Mazeriis cum decimis et primitiis, cum ingressibus et egressibus et cum omnibus sibi pertinentibus, que sic affrontat: a parte occidentali, confrontat in Garonna Mortua, ab oriente in condamina Sancti Saturnini, hanc condaminam sicut predictis confrontationibus includitur donamus et concedimus tibi Bertrando abbati Grandissilve et fratribus eiusdem loci, ut libere et quiete habeatis et possideatis perpetuo iure. Et ego prescriptus abbas et nos fratres . . . concedimus vobis partem et societatem in omnibus benefitiis spiritualibus. . . . Et si vos ad religionem veniatis recipiemus vos sicut fratres. Huius rei sunt testes: Raimundus Willelmi de Marcafava, Amelius de Montauros, Jordanus de Caraman, Arnaldus Pontii, Arnaldus de Alzacamba."

the west, and the land already conveyed by Arnold Helye to the south.[66] Bernard Pilgrim in turn gave his land located to the north of that belonging to Peter of Bagnols, and so on.[67] Such examples, often found side by side in the cartularies, could be multiplied, but this group of conveyances is particularly interesting since it concerns land which had presumably only recently been reclaimed from the Garonne.[68]

In most cases there is only a vague difference between the compacting activity already discussed and the second type of consolidation: the reassembling of rights from all owners on individual holdings. This second process of consolidation in which both lords and peasants gave up rights to the same holding or holdings might be called "vertical compacting" as opposed to the "horizontal compacting" just shown for Grandselve. This reconstitution of ownership involved the gradual redemption of all outstanding rights and claims to "ownership" on any specific field, meadow, farm, village or other property over which the order already held rights. By this process the order ousted all other owners from a particular property so that the monks controlled all rights there, and thereby became both lords and peasants, not to mention tithe-holders, owners of *feudum, allodium,* and so on. Such reassembling of rights to various levels of ownership is clearly implied in documents having clauses in which lords ceded rights to the monks of *feudum* or *vicaria* in areas where an abbey already held other rights like *dominium,* or in those documents in which tithes were given to the monks in areas where their granges had replaced the existing tenant-cultivators.[69] This vertical consolidation of rights or reconstitution of total ownership is explicit in examples such as the "donation" to Berdoues by William of Laseran of land at Lasmeades in the territory of Lafite for thirty *solidi*, and a corresponding "gift" of the *casal* of Lasmeades at Lafite

---

[66] Ibid., fol. 3v (n.d.): "Sit notum omnibus quod Ego Petrus de Baniols, et ego Guillelmus filius eius, per nos . . . donamus et concedimus in perpetuum domino Deo et Beate Marie Grandissilve et vobis fratribus eiusdem loci presentibus et futuris, quicquid habemus et habere debemus quoquomodo in terra de Mazeriis, que sic terminatur: ab oriente confrontat in honore Sancti Saturnini, a meridie confrontat in terra quam dedit vobis Arnaldus Helyias, ab occidente in Garonna Mortua, a septemtrione affrontat in terra Bernardi Peregrini. Et propter hoc dedistis nobis . . . vii. sol. Morlanenses. Huius rei sunt testes: Maurinus Capellanus, Pontius Cavallarius, Raimundus Miro, Raimundus Bover, Pontius filius Maurini, Bernardus Peregrinus."

[67] Ibid., fol. 3r (1152) "M.C.L.II. . . . Bernardus Peregrinus . . . vendo et dono in perpetuum domino Deo et Beate Marie Grandissilve et fratribus eiusdem loci . . . illam terram quam habeo iuxta Garonnam Mortuam, que afrontat ab occidente in eadem Garonna, a meridie afrontat in terra Petri de Baniols, ab oriente in honore Sancti Saturnini, a septemtrione in terra Pontiis Cavaler, . . . et dedistis michi propter hoc . . . vi. sol. Morlanenses. . . . Testes: Maurinus Capellanus, Pontius Cavallarius, Raimundus Miro, Raimundus de Bover, Pontius filius Maurini, Petrus de Baniols."

[68] These properties had earlier been held by the church of St. Sernin and seem to have formed part of a *sauveté* founded at Mazerières by St. Sernin in concert with the eleventh-century viscounts of Toulouse; see Higounet, "L'occupation," 301–30.

[69] Rodez, Soc. des lettres, "Cart. d'Iz et Bougaunes," nos. 16 (1168), 27 (1171), 43 (1179), 126 (1179), and 151 (1194), *Cart. de Bonneval*, no. 103 (n.d.), or *Cart. de Silvanès*, no. 501 (1198).

by Bonetus of Lasmeades, "who was accustomed to be tenant there," for fifteen *solidi*, or that of Bernarda of Sirac and her associates who gave up land in the parish of St. Martin as well as the *servitia* owed them by peasants there, or that of Bernard of Montaut and his family who gave Grandselve all their rights over men and their progeny living at Brugale and Montpré.[70]

It is often impossible to distinguish the two types of compacting. Both are exhibited in a slightly different kind of description found in the Villelongue documents, which gives virtually identical boundaries for properties in the *villa* of Villelongue, although here the documents do not say explicitly that the holdings were adjoining those already owned by the monks. The territory within the *villa* conveyed to the monks is described using perambulations, in this case citing the prevailing winds to denote directions, as "the honor held in the *villa* of St. John of Villelongue, bounded by the territory of Canavellas and of St. John of Valségier on the east [as the charter says *ab altano*], in the south by the territory of St. James of Villevalerian, in the west [*a circio*] by the territory of St. Martin le Vieux, and in the north by the territory of St. Mary of Varnassonne and by the woods [*silva*] of St. Benedict."[71] Here, the pattern is one of repeated conveyances within a certain territory from a whole series of individuals holding rights there. Although the charters do not explicitly show the acquisition of adjoining holdings, the reassembling process is implicit in such a group of conveyances. They include parallel "donations" by individuals at an equivalent level of the social hierarchy; all such "donors" on that level were paid comparable amounts for their land.[72] This series of conveyances shows not only the monks' compacting or consolidation of land rights from "donors" at the level of lordship, but also reveals that peasants were displaced. In this case, the peasants were removed when their lords (donors of rights to the Cistercians) gave them land elsewhere, in order to allow the monks to construct an abbey in the *villa* of Villelongue (to which the community was moved and from which it took its permanent name).[73]

---

[70] *Cart. de Berdoues*, nos. 500 (n.d.) and 506 (n.d.), and *Cart. de Gimont* III, no. 30 (1172): "insuper absolverunt servicium hominibus illius terre quod inde eis facere debebant"; ibid., III, nos. 10 and 11 (1162), the complementary grants by lord and peasants: "Quod Deusaida de Lombirag donavit, Bernardo abbati suam culturam que est juxta terram Raimundi d'Antincamp, et mandavit Donato de Cornelac ut predictam culturam monstraret et determinaret fratribus Gemundi. . . . M.C.L.XII."; and "Quod Bonus de Perissan et Giralda, uxor ejus, et Raimundus filius eorum, absolverunt Bernardo abbati totum hoc quod habebat in supradictam culturam quam dederat Gemundi habitatoribus Deusaida de Lombirag. . . . M.C.L.XII." For Grandselve see Paris, B. N. Coll. Doat, vol. 77, fol. 189 (1209).

[71] Paris, B. N., Coll. Doat, vol. 70, Villelongue, no. 1 (1165) fol. 1: "ab altano in terminio de Canavellas et terminio S. Johannis Vallis Sigerii, a meridie in terminio S. Jacobi de Villavaleriano, a circio in terminio S. Martini veteris, ab aquilone in terminio S. Mariae Varnassone et in silva S. Benedicti."

[72] Each was paid two or three hundred *solidi* for their conveyances which were made at the express wish of Lord Isarn Jourdain of Saissac: ibid., nos. 26-33 (all dated 1165), fols. 52-75.

[73] Villelongue's site change from Campagnes occurred ca. 1171; see ibid., no. 46 (1171), fol. 100, the earliest charter which refers to that community as Villelongue.

Since the *silva* of the charter descriptions was replaced by a *villeneuve*, it was probably into that woods of St. Benedict that the previous tenants of the *villa* were moved and it was probably those peasants who provided the manpower for its clearance and conversion to a new village.[74] Similarly, the extensive documents and sums expended by Valmagne for properties in or adjoining the *villa* of Vairac demonstrate consolidation, even if it is not always possible to show that all holdings were adjacent, or that multiple rights over individual holdings had to be acquired.[75]

In most cases, the process of compacting of strips and other holdings and that of repurchasing of ownership over each individual holding would have taken place concurrently, and in most cases, this consolidation was even less explicit than in that of Villelongue. In conveyances such as those to the Gascon abbey of Gimont concerning its grange of Hour, numerous documents mention land and land rights all in identical terms and all identified as being in a relatively small number of places: Manneville, Ardichol, or the parish of St. Martin of Toget.[76] These charters suggest that the reconstitution of holdings in a contiguous area and acquisition of rights over the same holdings from a number of individuals were under way. Just the fact that there were many conveyances to the monks all concerning the same territory or parish implies this reconstitution of ownership. For Berdoues's holdings in the former *villa* of Esparciac, for instance, multiple charters record conveyances to the abbey of numerous field strips (*culturae*), and show very easily that compacting was taking place, although only rarely can common boundaries be detected in the charter descriptions.[77] Although not all holdings mentioned can be proved to have been adjacent, the implication of consolidation is strong. A similar result can be seen in Berdoues's acquisitions in the parish of Durfort, which adjoined its grange of Cuelas, although there the consolidation seems to have been agonizingly

---

[74] Ibid., no. 49 (1178), has a *villeneuve* instead of the *silva* of St. Benedict in its description. Similarly, charters from the Silvanès cartulary assume that peasants would fill up uninhabited forest after Cistercian settlement, see *Cart. de Silvanès*, no. 381 (1147): "Nichilominus autem hoc attendendum est quod predictum territorium seu pascuam ad animalibus aliorum hominum defendere non habetis, et agriculturam ibi exercere volentes prohibere non debetis." Similarly, mortgagers of land to Berdoues retained rights to plow in the noval lands there, *Cart. de Berdoues*, no. 499 (1222): "Notum sit quod Gassias Ciche et Ramundus Hospitale et Willelmus Sanz de Lisos, bona fide miserunt in pignus Gillelmo, abbati et omni conventui Berdonarum totam terram cultam et incultam quam habebant a Lafite, ante Baisam et retro. Terminus redimendi hujus pignoris est de martro usque ad sanctum Joannem et non debetur despignorare pro alio aliquo homine hec predictum pignus et R. Hospitale potest arare, si voluerit in predicta terra aratro et hoc in novalibus tantum. . . . M.CC.XX.II."

[75] See for example, Montpellier, A. D. Hérault, film, "Cart. de Valmagne," vol. I, fols. 100r to 135v, and Vol. II, fols. 131r to 146v.

[76] *Cart. de Gimont* III, passim.

[77] Two *culturae* at Esparciac, *Cart. de Berdoues*, nos. 409 (1119) and 415 (1152); land near the grange of Esparciac, ibid., no. 381 (1164); land near Esparciac, ibid., no. 378 (1180); land at Bezmaux, ibid., no. 410 (1183); land at Esparciac, ibid., no. 424 (1190); land at Esparciac, ibid., no. 300 (n.d.); land at Esparciac mortgaged to Berdoues, ibid., no. 434 (n.d.); all rights in the territory of Esparciac between the Soussonne and the Pogge Rivers, ibid., no. 431 (1200).

slow. In 1211, the newly reconstructed parish church of Durfort was granted to the monks so that services there could be restored.[78] Then Berdoues gradually came to hold land nearby. For example, at Lafite four *casaux* had been acquired by 1237, but the fifth came to the monks only in 1255, more than a century after acquisitions for the grange of Cuelas had begun.[79]

Both vertical and horizontal consolidation can be detected in Silvanès's acquisitions for its grange of Promillac, which was located not far from that abbey's site in the southern Rouergue. For Promillac, the monks acquired a diversity of rights from many owners and of many sizes. Some of these were strips and small plots which had been parts of an old estate at Laurs, long since fragmented into tiny holdings in a variety of hands.[80] Also included in acquisitions for Promillac were tenures in the parish of Serruz where Silvanès had been granted the parish church by the bishop of Rodez.[81] There were also numerous gardens, vines, meadows, and riparian rights along the Dourdou River where the new grange would be built, as well as rights to detached *mansi* outside the original estate of Laurs and holdings in the neighboring *villa* of Magdarz.[82] Rights were also conceded to Silvanès for the construction of a mill at Promillac with two wheels, one to be used for fulling and one for grinding, with fishponds and adjoining irrigation channels for making hay meadows.[83] Finally, at Promillac, peasants owing services and rents were induced to cede their claims to the monks of Silvanès in return for remission of their obligations to the lord and promises of admission into the abbey.[84]

Such examples of reorganization and consolidation of land are frequently found in charters for Cistercian abbeys from throughout southern France, and are recognizable because those charters tend to be organized into cartularies by granges. Indeed, the cartulary organization itself is evidence that such consolidation was anticipated by the monks. Indeed, wherever multiple land acquisitions within a *villa* or territory are found in the doc-

---

[78] *Cart. de Berdoues*, no. 459 (1211): "Sciendum est quod Arnaldus Gillelmus de Lobcascha, in illa infirmitate de qua mortuus est positus in suo bono sensu et in sua bona memoria, amore Dei et remissione omnium peccatorum suorum, bono animo et bona voluntate, bona fide et absque ulla retentione quam ibi non fecit per se et per omnes successores suos presentes et futuros, donavit et reliquit in elemosinam Deo et beate Marie Berdonarum, Gillelmo abbati et conventui ejusdem loci presenti et futuro ecclesiam Sancti Stephani, que est juxta castellum de Durfort, cujus ecclesie sua propria jura dompnus Bernardus, auxitane ecclesie archiepiscopus, amore Dei ob restaurationem ipsius ecclesie, quam ille una cum fratre suo Stephano suis sumptibus et expensis restauraverat, . . . sicut jam diximus, donavit et reliquit eam fratribus Berdonarum ut sit illorum propria et libere et quiete teneant illam et accipiant decimas et primicias et quidquid ad illam pertinet. . . . Rogavit etiam predictus Arnaldus Gillelmus predictum Berdonarum abbatem quatenus pro amore ipsius duos filios clericos quos habebat in domo sua pro monachis reciperet. . . . M.CC.X.I."

[79] Ibid., nos. 505 (1237) and 508 (1255).

[80] *Cart. de Silvanès*, nos. 213 (1137) and 177 (1153).

[81] Ibid., 186 (1140), 195 (1143), 176 (1149), 177 (1153), 198 (1157), 216 (1160), and 231 (1164).

[82] Ibid., nos. 175 (1139) and 209 (1156).

[83] Ibid., nos. 210 (1159) and 192 (1162); discussed in chapter 4, pp. 88–89.

[84] Ibid., nos. 220 (1162) and 224 (1163).

uments, consolidation of land, both by compacting of holdings and reconstitution of all levels of ownership, was probably underway. Such consistent evidence of land amelioration suggests that this slow, careful repurchase was beneficial to the order's needs. It is certainly the best available explanation of how the order transformed parts of the rural countryside into the granges which were necessary to its economic program.

The gradual process of Cistercian acquisition of all rights within a territory included the purchase of tithes there. This is demonstrated in such agreements as that discussed in chapter 2 concerning the disposition of tithes in the parish of Corronzanges which was acquired by Bonneval.[85] Although in theory, Cistercian agriculture was exempt from the payment of tithes, only from tithes on their animals did Cistercians in southern France ever seem to have been automatically exempt (and after 1200 there were even exceptions to that). In southern France, where tithes had become attached to productive land and were bought and sold on the real estate market like any other rights to land, the order's tithe exemption could in fact only be exercised by repurchase of tithes from those into whose hands they had fallen.[86] This process of repurchase or acquisition of tithes by the monks, generally from the laity who had acquired them in previous centuries, was another aspect of the vertical consolidation of rights over land, although there were some differences from other land acquisition.

The Cistercian tithe exemption derived from a papal privilege granted in 1132 by Innocent II to Cîteaux and its congregation, which was often reiterated for individual houses of the order in papal privileges that also confirmed their property holdings.[87] This privilege regarding tithes specified that the order pay no tithe on the produce of lands which were worked under Cistercian management, or on its stock-raising.[88] Confirmations of

---

[85] *Cart. de Bonneval*, no. 23 (1176). See p. 18 above.

[86] On this transformation of tithes from personal to predial, see Catherine E. Boyd, *Tithes and Parishes in Medieval Italy* (Ithaca: Cornell University Press, 1952), 139–40. This situation seems to have been fairly widespread; Constable refers to transfers of tithes as "sales of property," *Monastic Tithes*, 272; this was certainly true in southern France. In that region, however, animals were still in the twelfth century considered under the personal conception of tithes; see chapter V. Examples of tithe conveyances are: *Cart. de Gimont*, IV, no. 89 (n.d.); or *Cart. de Bonneval*, no. 28D (n.d.), or *Cart. de Bonnecombe*, no. 286 A (1205): "M.CC.V. Ego Willermus Seinnorellz, per me . . . dono et concedo Deo et Beate Marie Bonecumbe . . . , alodium mansi Gervaissenc, qui alio nomine Ugonencz dicitur, cum omnibus pertinenciis suis et omne quod querebam vel querere quomodocumque poteram in alodio mansi Cairelencz sive in decimis istorum mansorum, . . . Et propter hoc accepi a vobis jamdictis fratribus in caritate XX solidos, quos integro numero me accepisse affirmo. . . . Et Bernardus Ugonis persolvit denarios ipsi Willermo in ospicio Bonecumbe."

[87] Constable, *Monastic Tithes*, 241–45.

[88] See *Cart. de Silvanès*, no. 6 (1140), for a privilege of Innocent II: "Sanccimus etiam ut de laboribus, quos propriis manibus aut sumptibus colitis, sive de nutrimentis vestrorum animalium, nullus omnino clericus sive laicus decimas, oblationes sive primicias, seu de terra vestra terraticum a vobis exigere audeat" and *Cart. de Bonnecombe*, no. 15 (1172): "Sane laborum vestrorum, quos propriis manibus aut sumptibus colitis sive de nutrimentis vestrorum animalium nullis a vobis decimas vel primicias sive oblationes, quas vulgo proferentias vocant,

the exemption, even though they often involved conveyance of the church in question to the monks who had repurchased its tithes, were made by local bishops with little complaint. This was probably because those tithes and churches had long since passed from episcopal control, and conveyance of those churches to the monks would actually have guaranteed revenues which were otherwise not collectible.[89] For example, the bishops and canons of the cathedral of Rodez eventually conveyed a group of parish churches to the Cistercian abbey of Bonnecombe in return for small annual rents.[90] Other members of the secular clergy also sometimes sold tithes to the Cistercians for nominal amounts; for example, a priest in the diocese of Béziers sold the tithes which he had recently repurchased from lay owners to the monks of Valmagne, for exactly the sum which he had dispersed in their acquisition.[91] Where tithe acquisition differed from other land acquisitions, if at all, was that recuperation of tithes from the laity was probably made simpler for the Cistercians because they were armed with their papal privilege as well as other canons condemning lay ownership of tithes.[92] The laity, threatened with damnation of souls for simony (the sin of trafficking in church property), thus willingly conceded tithes and churches to the Cistercians in return for cash gifts or annuities from the monks.[93]

Tithe repurchase from other monks and the regular clergy was more difficult. Grave controversy arose when other monasteries or religious held tithes and churches in parishes where the Cistercians were creating their granges. In general, when such conflicts arose, the order had either to agree to pay a fixed annual rent in lieu of tithes to the other religious corporation, or to cede property to the claimants in return for freedom from tithes. For example, Bonnecombe eventually agreed to pay a fixed annual rent instead of tithes to the Templars for lands which the Cistercians cultivated in the

---

exigere presumat." See also Montpellier, A. D. Hérault, film, "Cart. de Valmagne," vol. I, fol. 1v (1147), for confirmation of tithe exemption to Valmagne before Cistercian incorporation, and fol. 3r (1162) for Alexander III's confirmation of the tithe exemption for Valmagne, and its mother-house, Bonnevaux.

[89] For example, the confirmation of tithes and first fruits on lands which Cistercians of Grandselve worked with their own hands or at their own expense in the parishes of St. Sulpice of Boulhac and St. Sernin of Ricancelle from the bishop of Toulouse: Toulouse, A. D. Haute-Garonne 108 H 56, fol. 97 (1163). Loss of episcopal control of tithes and parish churches is apparent from the numerous concessions of tithes to the order by the laity. Of 151 churches conveyed to the Cistercians in southern France, only forty were conveyed to the monks by bishops; see Constance H. Berman, "Cistercian Development and the Order's Acquisition of Churches and Tithes in Southern France," *Revue bénédictine* 91 (1981): 193–203.

[90] *Cart. de Bonnecombe*, nos. 1–6. These seem to be rights of some value at least to the prestige of bishop and chapter since nominal rents are paid for them.

[91] Montpellier, A. D. Hérault, film, "Cart. de Valmagne," vol. 2, fol. 116r ff. (1191).

[92] See ibid., fols 1r-6v, pages on which Valmagne's monks had copied a number of canons concerning tithes into their cartulary. Calers also had several parchments containing long canons on tithes: Toulouse: A. D. Haute-Garonne, 108H Calers, non coté.

[93] Tithes or churches conveyed for cash are found widely in the charters; for example, *Cart. de Gimont* II, no. 43 (1162) and no. 136 (1164). More often, tithes and first fruits are conveyed with other rights to land, as ibid., nos. 187 (1162), 3 (1163), and 136 (1164).

parishes of Iz, Ampiac, and Limosa.[94] In the southern Rouergue, Silvanès, having similar difficulties over tithes in the parish of Prugnes, came to an agreement in which a tenth of its land held there was ceded to the Hospitallers there in lieu of tithes.[95] As already mentioned, in order to compensate that church for tithes lost, Bonneval paid the parish church of Corronzanges a tenth of the number of *solidi* which the monks dispersed in land acquisitions in its parish.[96]

Such arbitrated settlements of disputes between Cistercians and other religious groups over tithes show the lengths to which the monks went in creating the unencumbered properties on which they wished to practice their "grange agriculture," and how Cistercian land consolidation encompassed even the rights to tithes from which the order was supposedly exempt. By the late twelfth century, the advantages of tithe-exempt agriculture were apparent. Growing complaints about Cistercian exemption were made by churchmen, who resented Cistercian advantages which allowed them to sell at lower cost, and who would have preferred to have had tithes restored to them rather than to the monks. This controversy caused the Fourth Lateran Council of 1215 to limit Cistercian tithe exemption to those lands already in hand at the time of the Council and for later acquisitions only to lands which could be considered "noval," that is, lands which the order had itself cleared or drained and on which tithes had never before been assessed. In theory, the exemption from tithes on Cistercian stock-raising was also maintained, although there is less clarity on this point.[97] The new ruling however, had little effect on Cistercians in southern France, where land had already been acquired at most grange-sites by 1215 and where the exemption had always been exercised only after outright purchase of tithes.

Acquisition of tithes has sometimes been used as a measure of Cistercian "decadence" because it led to the order's acquisition of such "forbidden" properties as tithes and churches. In fact in southern France it was simply another facet of the Cistercian land consolidation program. Moreover, tithes and churches acquired by the order meant that tithes and churches had been restored to ecclesiastical ownership, albeit monastic ownership, an important part of the Gregorian reform. Such churches and tithes were almost never acquired except in the context of reconstitution of agricultural rights,[98] and where tithes were among the rights to land acquired by the

---

[94] Rodez, Soc. des lettres, Bonnecombe, "Cart. d'Iz et Bougaunes," nos. 251 (1154), 114 (1194).

[95] *Cart. de Silvanès*, nos. 170 (1154), 174 (1165).

[96] *Cart. de Bonneval*, no. 23 (1176).

[97] Constable, *Monastic Tithes*, 270–306. None of the complaints he cites refers to southern France specifically; on the subsequent controversy, particularly with regard to the need for clarification on the tithes over "nutrimentum" for Cistercian stock-raising, see Mahn, *L'ordre cistercien*, 112–15.

[98] My conclusion discussed in Berman, "Cistercian Tithes," 202–203, that tithes and churches were not sought for their revenues *per se* but because they were needed for the order's agricultural program contradicts earlier interpretations such as that of Hill, *English Cistercian Monasteries*, 109–12, who presents tithe acquisitions as an index of decadence.

monks, they are indicative of completeness in the order's land consolidation and grange development.

Finally, the reassembling or repurchasing of dispersed rights over land by Cistercians in southern France included the acquisition of rights to cultivate from the men and women who had traditionally tenanted holdings coming into Cistercian ownership. The gradual removal of such peasants having heritable claims to land which the monks desired for their granges, allowed abbeys to institute the direct cultivation for which they were famous and for which the order was tithe-exempt. Since peasants tended to be conveyed with rights to land when lordship was transferred, the order often acquired dependent tenants whether wanted or not. The presence of such peasants on land coming into Cistercian ownership caused difficulties in land acquisition which had not arisen on earlier Benedictine estates even when they were expanding. Earlier Benedictine houses, like Cluny, had acquired land with its existing tenants and proceeded to collect rents, dues, and services, without any change or reorganization of the holdings; as tenancies fell vacant, they were let again; when land was transferred, tenants were transferred with it.[99] For Cistercians, however, such existing tenants complicated the transformation from *villa* to grange. Creating Cistercian granges involved a total reorganization of the land-tenure system—a change in the actual workings of the rural countryside which was as complete a transformation as any clearance of forest or drainage of swamp would have been. The program of the new order, advocating self-sufficiency for the monks, assumed that foundations would be in uninhabited areas of wilderness where land would be cultivated by members of the community—monks and lay brothers. It provided no guidance for the situation in which land was acquired with tenants on it. The ideals of the order also assumed that recruits would be drawn to the order and to its lay brotherhood out of religious zeal, like those monks who had come with Bernard. It did not consider that it might be necessary to seek *conversi* from among the previous tenants on land coming into Cistercian hands, or that peasants were not free to become *conversi* at will except when Cistercians became their lords.[100] Thus, in the transformation of previously cultivated lands into Cistercian granges, the order's ideals were challenged, both by the existence and claims of peasant men and women to hereditary rights to

---

[99] On earlier monastic land administration, see Marc Bloch, *Seigneurie française et manoir anglais* (Paris: A. Colin, 1967), 23–47; Duby, *Rural Economy*, 197–220; Elisabeth Magnou-Nortier, *La Société laïque et l'église dans la province ecclésiastique de Narbonne* (Toulouse: Université de Toulouse, 1974), 153ff. and 539–40; Duby, "Le budget de l'abbaye de Cluny," *Hommes et structures du Moyen Age. Recueil d'articles* (Paris: Mouton, 1973), 61–79.

[100] On the program outlined in the statutes, see Lekai, "Ideal and Reality," 4–7, and Roehl, "Plan," 83–90. It is not clear to what extent either of them has considered whether the plan as outlined in the order's earliest statutes would really have worked in the rural conditions of twelfth-century Europe. There seem to be internal inconsistencies, for instance, with regard to *conversi*, as discussed below.

cultivate, as well as by the general economic dependency of most peasants on their lords.

Transforming peasants into *conversi* was only one of several ways in which Cistercians rid their lands of previous occupants, but it was an extremely important aspect of the acquisition process because of the importance of *conversi* to Cistercian grange agriculture.[101] Although in the early years the monks worked alongside those lay brothers in the fields, increasingly the larger share of the fieldwork, particularly on the satellite farms, must have fallen to the lay brothers. The *conversi* were able-bodied, celibate men, who entered a monastery to be farm laborers, but farm laborers with a difference, since they were free from family responsibilities, from obligations to lords, and from many of the insecurities of the peasantry. Moreover, the status of *conversus* in a new Cistercian abbey in the twelfth century, when it was little different from that of a monk and when even knights entered the order as *conversi*, was an enviable position. For many twelfth-century peasants, the status of *conversus* would have been desirable not only because of its economic security, but because it fulfilled religious aspirations which had earlier been denied. Only in the 1180s or 1190s did *conversi* begin to become second-class members of monastic communities, and the real decline in *conversi* status was probably only somewhat later, in the mid-thirteenth century.[102] For most peasants, however, the opportunity to become *conversi* only existed if the Cistercians became their landlords, for there is no indication that most other lords would have allowed this dependent labor force to desert holdings at will to become *conversi* on Cistercian granges.[103] Moreover, few peasants could have fled previous

---

[101] Discussion of the lay brothers or *conversi* of the Cistercian order abounds, but confusion remains. Throughout the twelfth century, class lines between *conversi* and monks seem to have remained fluid, as is clear for southern France from examples cited above in n. 25. Moreover, there has been a mistaken attempt by historians to equate those entering a house *ad conversionem* with the *conversi*—two phenomena which were often, although not always, distinct. Giles Constable has discussed the problem with regard to Cluny's members in "*Famuli* and *Conversi* at Cluny," *Revue bénédictine* 83 (1973): 326–50. James S. Donnelly, *The Decline of the Medieval Cistercian Lay-Brotherhood* (New York: Fordham University Press, 1949), does not address the twelfth-century terminology or the origins of lay brothers. It is symptomatic of the continued confusion on the problem that there is no more recent article on the origins of lay brothers, or the distinctions between lay brothers and monks during the twelfth century than K. Hallinger's "Woher kommen die Laienbruder?" *Analecta Cisterciensia* 12 (1956): 1–104. Hallinger does not use cartularies as a source, but has consulted necrologies for statistics on the incidence of the term "*conversus*." A thorough quantitative study of the problem of *conversi* origins and of the distinctions between *conversi* and monks, using the evidence of local documents, needs to be undertaken. Furthermore, the problem of where *conversi* could have been found on the frontiers of Europe must be considered.

[102] The first time that class distinctions are mentioned in the order's legislation is in 1188, "Nobiles laici venientes ad monasterium non fiant conversi sed monachi": *Statuta*, ed. Canivez, vol. 1 (1188), no. 8.

[103] On this dependence, see Paul Ourliac, "Le servage à Toulouse aux XII⁰ et XIII⁰ siècles," *Economies et sociétés au Moyen Age: Mélanges offerts à Edouard Perroy* (Paris, 1973), 249–61. On land holders retaining their peasants even when land is conveyed to the monks, see Paris, B. N., Coll. Doat, vol. 70, Villelongue, nos. 34, 35, 36 (1165), fols. 74r–79r, etc.

lords to become *conversi;* for those willing to embark on fugitive lives there were undoubtedly better alternatives in the towns, on the Spanish frontier, or following Crusaders, than in neighboring Cistercian houses.[104]

Potential *conversi* did not just arrive at the gates to be admitted. Former tenants of lands coming into Cistercian ownership seem to have been the main source of such *conversi* recruits. If those tenants had not been there, it would have been much more difficult for the order to recruit able *conversi;* indeed, once Cistercians had lordship over dependent tenants, they could easily encourage them to become lay brothers. Only when lords were selling all rights to an area, including their tenants, could peasants have been transformed into lay brothers. Indeed, recruitment even of hired laborers must have often been difficult for the early order. Legal controversies arose in which Cistercian houses were forced to respect the rights of other consumers to manpower in the community; a much damaged and badly transcribed copy of a document concerning Belleperche, for instance, suggests that the monks of that house came into conflict with nearby nuns over hiring local laborers during the harvest.[105]

Such assertions about the relationship between land acquisition and the recruitment of *conversi* are not based on surmise. Many charters document the concession of rights to enter Cistercian houses as *conversi*, to peasants who had previously been tenants on land coming into Cistercian hands, generally in return for those peasants' rights to cultivate. Peasants conveying such tenurial rights to the order were either admitted immediately as *conversi* of the Cistercian house, or were promised such admission in the future.[106] Although peasants might be allowed to stay on a tenure for a lifetime, with sons entering as *conversi*, or until they were too old to work lands themselves, most often such promises to be accepted as *conversi* were made with a time limit: for instance, one man was required to enter before the next Christmas, another within three years.[107] Moreover, particularly in the Gascon charters, it is often specifically required that the individual who entered as a lay-brother had to be well and able-bodied—with all his limbs intact.[108] This affirms that farm laborers and not dependents were sought. Probably peasants were urged to enter abbeys as *conversi*—and transfer their lands to the monks as quickly as possible—so that grange creation could go forward and because they were needed as laborers. Cases from the Rouergue showing explicitly the transfer of rights to cultivate (*pagesia*) along with permission for peasants to enter a Cistercian house as lay broth-

---

[104] On such migration to towns, see Charles Higounet, "Le peuplement de Toulouse au XIIe siècle," *Annales du Midi* 55 (1943): 489–98.

[105] Toulouse, A. Mun., MS 342, Belleperche.

[106] For example, Pons Matreville gave land to Boulbonne and was granted permission to enter as *frater* or monk: Paris, B. N. Coll. Doat, vol. 84, Boulbonne, no. 169 (1230).

[107] *Cart. de Gimont*, I, no. 68 (1180), VI, no. 61 (1173); *Cart. de Berdoues*, no. 494 (n.d.) and no. 7 (1161).

[108] The Berdoues charters are very explicit about this: *Cart. de Berdoues*, nos. 275 (1183), 604 (n.d.) and no. 7 (1161).

ers have already been mentioned. At Vareilles and elsewhere, Bonnecombe promised admission as *conversi* to peasants in return for *pagesia* rights.[109] Examples from the cartulary of Silvanès record men from Promillac giving up land in one act and appearing in other charters as *conversi*, or men from Sallelas and Landes ceding property in those places in return for promises of admission as *conversi*.[110]

If peasants were available to be recruited into Cistercian houses only in settled areas where Cistercians were replacing their lords, then Cistercian houses had to acquire at least some of their properties in areas of previous settlement. A peasant population that would be attracted as *conversi*, or as skilled farm laborers for seasonal hiring, could not be found in previously unsettled areas or the frontiers of that period (except perhaps in Spain where Moslem laborers may well have been used).[111] Such a link between *conversi* recruitment and monastic land-acquisition suggests the need for some nearby peasant community from which to recruit laborers, and may explain why Cistercian foundations made on the real fringes of Europe were rarely as successful as those having contact with a settled agricultural populace.[112] If it was only through land acquisition that the order had the opportunity to recruit peasants to become lay brothers, then the issue of labor recruitment may have forced the order to found its houses and granges in areas of previously settled and cultivated lands.

It was these *conversi* laborers, assisted by the monks or by hired agricultural workers, who did the primary work on the granges and it was through their efforts that grange agriculture made Cistercians rich. Eventually, however, the order would discover the disadvantage of such a labor force of celibate men: it did not replace itself from generation to generation as had the peasant farm family. Particularly after 1250 when the order's land acquisitions ceased, *conversi* recruitment became a problem. Such difficulties with *conversi* recruitment in the thirteenth century, often accompanied by an increase in the incidence of *conversi* revolts throughout Europe, have long been recognized, and the decline in *conversi* numbers in the mid-thirteenth century has often been described as a "failure of religios-

---

[109] *Cart. de Bonnecombe*, nos. 264 H (1183), 268 C (1187), 272 C (1192), 273 C (1193).

[110] *Cart. de Silvanès*, nos. 138–40 (1159) for Landes and Sallelas, and no. 319 (1167) for Rouret and document quoted in chapter 2, n. 29. Similar examples are found in *Cart. de Berdoues*, no. 242 (1150) and *Cart. de Gimont*, III, no. 2bis (1159). Sometimes the previous status of an entrant is obscure, as when Peter Rainaldi entered Silvanès as a *conversus*, making arrangements that his son enter as a monk; *Cart. de Silvanès*, no. 202 (1158).

[111] On the difficulty of attracting lay brothers in Spain and the necessity of maintaining lay tenants, see Lawrence J. McCrank, "The Cistercians of Poblet as Landlords: Protection, Litigation, and Violence on the Medieval Catalan Frontier," *Cîteaux: comm. cist.* 27 (1976): 255–83, especially 274–75. No mention is made of Moslem peasants.

[112] On the difficulties of Cistercian settlement, for instance, in Norway, see Arne Odd Johnson, *De norske cistercienser-klostre, 1146–1264* (Oslo: Universitetsforlaget, 1977), especially 48–66. On similar difficulties in Danish houses on the Baltic, see McGuire, *Cistercians*, 151–55. In the Latin Kingdom of Jerusalem, Cistercian foundations disappeared very early; E. A. R. Brown, "The Cistercians in the Latin Empire of Constantinople and Greece, 1204," *Traditio* 14 (1958): 63–120.

ity."[113] If, however, *conversi* recruitment is understood to have been closely tied to land acquisition, the decline in *conversi* numbers can no longer be viewed as a failure of reform ideas. Instead, it was probably the natural result of aging and death among the original *conversi*, the beginnings of their dependency on the order, and the end of the land acquisitions to which *conversi* recruitment was linked.

Not all peasants whose lands were acquired by the Cistercians became *conversi*. It must be assumed that some continued to cultivate holdings held under previous lords—particularly vineyards, olive groves, or gardens requiring intensive labor. Tenants probably also often remained on isolated holdings which were never wholly integrated into the grange system.[114] In a few cases charters explicitly gave such peasants the right to stay on their tenures during a lifetime, with only the next generation displaced by monastic acquisition. For example, cultivators of land at las Clotas were allowed to stay on land which they had granted to Grandselve for as long as they lived; similarly Raymond Vidal retained for his lifetime an island in the Garonne near the lands which he had reclaimed along that river and which he had granted to Grandselve.[115] Elsewhere, the Grandselve charters show that services owed to previous lords by men whose tenancies came into Cistercian hands were transferred to the monks.[116] In such cases, peasants may have been legally free, but they were in fact attached to the land by economic obligations to pay rent. These economically dependent peasants got an opportunity to divest themselves of such obligations by conveying their tenurial rights when Cistercians acquired lordship over

---

[113] The crisis of *conversi* recruitment is described by Donnelly, *Decline*, 38–60; both he and Lekai attribute it to a failure of religiosity; see Lekai, *The White Monks*, 339–45, who remarks on this problem of the "strength of religious motivation." Neither sees any relationship to the end of expansion, nor does Roehl consider the problem of *conversi* replacement in discussing the advantages of *conversi* labor: "Plan," 87–95. See *Statuta*, ed. Canivez, 1190, no. 75; 1212, no. 52; 1238, no. 52, etc., for southern French *conversi* revolts.

[114] Many such holdings, however, were exchanged or leased. See *Cart. de Berdoues*, no. 65 (1211): "Sciendum est quod Gillelmus abbas Berdonarum et fratres ejusdem domus dederunt a fius [sic] Vitali Grillo de Serris dicto filio Ramundi Grilli de Serris et omnibus successoribus ejus presentibus et futuris quamdam vineam quam habebant a Serras que vinea fuit Oddonis de Gaviano qui fuit monacus prefate domus Berdonarum ut habeant et possideant illam libere et quiete sine omni contradictione et teneant illam bene condreitam in perpetuum, ita ut pro ipsa vinea annuatim in die et festivitate omnium sanctorum dent absque ulla contradictione et mora VIII denarios morl. oblias sacriste domus Berdonarum. . . . M.CC.X.I." Bonnecombe rented out half of a *mansus* at Saunnac: Rodez, Soc. des lettres, Bonnecombe, "Cart. de Moncan," no. 65 (1185) and Rodez, Soc. des lettres, Bonnecombe, "Cart. d'Iz et Bougaunes," no. 101 (1191); a field at Bogath: no. 242 (1202); and an early exchange, 1177, with the hospital of La Selve: Rodez, Soc. de lettres, Bonnecombe, "Cart. de Magrin", no. 20 (1177). Its grange of Bernac gave up rights in the *mansus* of Mocenes and men and women there: Rodez, A. D. Aveyron, 2H Bonnecombe, "Cart. de Bernac," no. 7 (1202); three *eminadas* elsewhere, ibid., no. 17 (1216); and vines at Sevalz, ibid., no. 49 (1209). Similarly, see the many contracts renting out vineyards for Grandselve or its Toulouse hospice: Toulouse, A. D. Haute-Garonne, 108 H Grandselve, liasses 1–40.

[115] Paris, B. N. Latin MS 11011, Grandselve, no. 263 (1180), and Latin MS 11008, Grandselve, no. 950 (1183).

[116] See, for example, the example cited in chapter 2, n. 32.

their holdings. These transactions allowed peasants to migrate to towns, move to other districts, or to use cash paid them for the piece of land sold to the monks to make improvements on other holdings, as well as giving them the option of becoming *conversi.*

Peasants were not always allowed to cut their ties to earlier lords even when their land came into Cistercian hands. In a few cases, lords specifically excluded the transfer of laborers when land was conveyed to the monks. In these cases, tenants were moved to new villages or other estates belonging to their lords when land was sold to the order. Thus, the documents for Grandselve record that tenant cultivators were promised new holdings elsewhere on their lord's land, after they had ceded their tenures to the Cistercians.[117] Similarly, in the Berdoues cartulary, a charter mentions that peasants had been given the option of staying on as tenants of the Cistercians or removing to new tenancies under their old lord.[118] An example of this removal of peasants cultivating land was already discussed: peasants from the *villa* of Villelongue were transferred, presumably to a nearby *villeneuve* in the 1160s when Cistercian monks began to acquire Villelongue. In that case, the indirect result of Cistercian acquisition of an established *villa* or village was probably the clearance of nearby forest by the *villa's* tenants, who were allowed to provide themselves new holdings. Despite such examples, the charters generally imply that the men and women living in the *villae* or on the farms conveyed to the Cistercians were treated as appurtenances to the lordship of those holdings; authority over such tenants was automatically conveyed to the monks along with ownership or *dominium* of the land. Once *dominium* was acquired, the Cistercians either transformed those tenants into *conversi,* or removed them from their old tenancies, usually by a simple purchase of their tenurial claims. Although the precise details of such removals are obscure, most existing peasants were displaced, as is clear from the evidence showing the institution of direct cultivation by Cistercian monks and lay brothers discussed below.

Occasionally, the removal by the Cistercians of peasants from land which they had tenanted or which those peasants had themselves cleared or reclaimed was made only under pressure. Two suggestive cases can be cited from Cistercian documents for southern France. The first is that of peasants who were urged for several years before they gave up the *finages* which they had cleared in the forest where the new house of Bonnefont was to be founded.[119] A second case concerns the murder of a servant or lay brother of the Cistercian house of Berdoues by men of a nearby village. That servant had been attempting to lay out grange boundaries for the monks when he was attacked and murdered. That it was the men of the

---

[117] Paris, B. N., Latin MS 9994, Grandselve, no. 662 (1167), and Latin MS 11008, Grandselve, no. 178 (1178).
[118] *Cart. de Berdoues,* no. 454 (1215).
[119] *Rec. de Bonnefont,* nos. 1 and 3 (1136–38).

village as a group who later made a gift to the monks in compensation for their crime suggests that they had formed an organized resistance to the monks' expansion. In this case, their violence was as unsuccessful as most peasant revolts have been, since in the end the monks received the land which they had desired.[120] This incident suggests that some of the so-called revolts of *conversi* in the thirteenth century might actually be viewed as revolts of peasants unwillingly converted to the Cistercian life during land acquisition. Unfortunately (with the possible exception of disputes over pasture, discussed in chapter 5) most resistance to monastic expansionism on the part of local lords or peasants would probably have gone unrecorded in the order's documents, so that it is impossible to know how welcome the planting of a Cistercian house may have been to the peasantry or to certain lords in many areas. The success of the order in southern France, however, denotes a generally positive reception by the neighboring communities.

The removal of existing peasants from land which the Cistercians desired to transform into granges and the transformation of those peasants into *conversi*, confirm the complexity and completeness of the land reorganization carried out by the Cistercians in southern France. Displacement of peasants along with other aspects of consolidation reflect how careful planning and purchasing made it possible for Cistercian houses in southern France to create a series of compact agricultural estates, despite the region's lack of untouched land. Such reorganization of land created granges on which the order's monks and lay brothers not only supported themselves, but produced agricultural surpluses. Such surpluses, sold in regional markets, provided the cash which was used for purchase of additional land, which in turn provided more surpluses. Thus agricultural growth continued.

The Cistercians were well suited to undertake the managerial tasks involved in such reorganization of land into granges, but the completeness of such reorganization is hard to measure. There are indications from a number of places that Cistercian consolidation did not always remove all trace of previous occupation from the landscape. The isolated *mansi* and villages incorporated by the monks in the twelfth and thirteenth centuries have often reappeared on modern maps.[121] It is in the same regions where

---

[120] *Cart. de Berdoues*, nos. 481 and 483 (n.d.) and no. 470 (1154): which tells us the murder and Berdoues's subsequent acquisition of the land: "Sciendum est quod Gauzion de Tornon et Petrus de Sobaian et homines de Ponsan contendebant cum fratribus Berdonarum de la terra del Pin, que terra est a Cuelas.Et cum diu in hac contentione perstitissent postea fecerunt accordium et statuerunt diem et exterminaverunt. Ex inde statuto die exterminabat predictam terram quidam homo de Guiseriz et fuit occisus ab eis qui erant ex parte de Ponsan. Unde Berdonarum fratres clamaverunt se ad archiepiscopum auxitanum et postea Gauzions et Petrus de Sobaian et Bernardus Ramundus filii de Na Gauzion decerunt accordium pro morte istius hominis pro se et pro omnibus successoribus suis presentibus et futuris, cum fratribus Berdonarum et reliquerunt et absolverunt predictis fratribus Berdonarum predictam terram."

[121] Particularly in the Rouergue. See areas around Magrin and LaFon where *mansi* of Cairo, Cairaguet, etc., still appear on modern maps.

such place-names are still found, however, that Cistercians continue even today to have the reputation for having created the largest, most valuable farms of the pre-Revolutionary period.[122] The development of such farms or granges had been envisioned and desired by early Cistercian legislators, but the careful planning and patience needed to reorganize fragmented holdings into such granges, including the need to displace existing peasants or convert them into *conversi*, were compromises between the order's ideals and actual conditions. The success of the adaptation is clear from the continued presence of Cistercian houses and their granges in southern France right up to the eighteenth century.

---

[122] The great size of these estates is apparent from as early as the fourteenth century when they became targets for marauding *routiers* during the Hundred Years' War: see Constance H. Berman, "Fortified Monastic Granges in the Rouergue," *The Medieval Castle: Romance and Reality*, ed. Kathryn Reyerson and Faye Powe (Dubuque, Iowa: Kendall, 1984), 124–46.

IV. THE PROFITS OF GRANGE AGRICULTURE

By consolidation, successful Cistercian houses of southern France gradually acquired the extensive lands which could be converted into agricultural units of a size larger than a whole parish or *villa*. These holdings were generally called granges. As has been shown, such granges were created by acquisitions which were closely tied to the order's recruitment of lay brothers. Using the services of such *conversi*, the monks instituted what might be termed the "new grange agriculture" for which the Cistercian order has become famous. This included the direct management of cultivation on large, uninterrupted expanses, the exemption of production on those lands from payment of tithes or other agricultural taxes, and the use of lay brothers and hired day workers in place of tenant farmers. The institution of such grange agriculture by the Cistercians at many sites throughout southern France is implied not only by the presence of those granges for centuries afterward but also in a number of indirect ways by the order's early charters.

The extensive reorganization of ownership and cultivation, which the creation of such granges in southern France represented, is a monument to Cistercian organizational skills, for by converting existing estates which were disorganized and fragmented, owing taxes to many and with few improvements or efficiencies, into "granges," the Cistercians cleared these holdings of a wealth of liabilities which must have consistently kept profits low for earlier owners. An assessment of the profitability of this Cistercian grange agriculture as it was instituted in twelfth- and thirteenth-century southern France can be made, but it must rest almost entirely on indirect information found in charters of land conveyance to the monks. There are no records of grange production, of gross income, or of profits from sales which might supplement the evidence of the surviving real estate records. While it is impossible to cite production figures or calculate actual yields, it is possible to show that Cistercian agriculture was more profitable than that which had preceded it. Existing records suggest that the regrouping of land, in and of itself, must have significantly increased the efficiency and profitability of agriculture. Production must have been improved considerably in comparison to that under earlier owners and cultivators, simply because such consolidation had been carried out.

Looking first at the grange itself—what comprised a "grange" varied. Nonetheless, it was always something more than solely a farmstead, although each grange had a complex of buildings at its nucleus. Perhaps a Cistercian grange is best described as an assemblage of fields or farms whose parts and whole varied considerably in size, but which were all organized into a single unit for agricultural production. This unit was generally administered by a lay brother, called the grange master or granger.[1] As is clear from recent studies of Cistercians in various parts of Europe, the grange was not a miniature monastery, nor were the Cistercians the only group developing granges at this time.[2] Moreover, the Cistercian grange was different from a deanery such as those found among the estates of certain Benedictine houses, or from a priory, although a few granges had their origins as priories.[3] Indeed, in theory the Cistercian grange did not have to have a church, although there were a number of exceptions in southern France. Such exceptions are found where an incorporated hermitage or priory had been reduced to a grange, or if a village with an existing church became a grange center; in such cases altars were maintained, but there is no evidence of the construction of new churches or altars at grange centers by the order in this region.[4]

No two granges could possibly be identical, even within the patrimony of a single abbey. Each grange differed from all others, varying in size, extent of consolidation, geographical conditions, profitability, and so on. Moreover, not only did the acreage included in such units, the number of laborers attached, and the number of animals kept vary from place to place, but at any grange these factors were constantly shifting as land was acquired and as staffing needs changed. In some places Cistercian grange centers (the actual farmyards) might develop into huge complexes, like that of Bonneval's grange of les Gallinières in the Rouergue.[5] In other cases the grange yard remained a simple group of house, wine cellar, and mill, as at Bonnecombe's "grange" of Bougaunes.[6] A number of such Cistercian grange steads or complexes still stand today in southern France, although

---

[1] On Cistercian granges, see Coburn Graves, "Medieval Cistercian Granges," *Studies in Medieval Culture*, ed. John R. Sommerfeldt (Kalamazoo, Mich.: Cistercian Publications, 1966), 2: 63–70, who discusses the origin of the term "grange"; Colin Platt, *The Monastic Grange in Medieval England: a Reassessment* (London: Macmillan, 1969), 46–49; R. A. Donkin, "The Cistercian Grange in England in the Twelfth and Thirteenth Centuries, with Special Reference to Yorkshire," *Studia monastica* 6 (1964): 95–144, idem, *Cistercians*, esp. 51–67; and Charles Higounet, *La Grange de Vaulerent, structure et exploitation d'un terroir cistercien de la plaine de France: XIIe-XVe siècle* (Paris: S.E.V.P.E.N., 1965).

[2] Platt, *Monastic Grange*, 71 ff., and Donkin, "Cistercian Grange," 100–101.

[3] Unlike the Cistercian granges, where only *conversi* were permanently stationed, both deaneries and priories often had monks sent there as permanent administrators. See in contrast, deaneries, as discussed by Penelope D. Johnson, *Prayer, Patronage and Power: the Abbey of la Trinité, Vendôme, 1032–1187* (New York: New York University Press, 1981), 53–56; and the priories of Cluny, Bede K. Lackner, *The Eleventh-Century Background to Cîteaux* (Washington, D.C.: Cistercian Publications, 1972), 69–70.

[4] For example, the church of the parish of St. Félix was undoubtedly maintained by Bonnecombe once St. Félix became a grange; see the conveyance in *Cart. de Bonnecombe*, no. 8 (1221), and n. 13 or n. 21 below. Similarly, Berdoues was granted at least one parish church near the site of its grange of Cuelas in order that it be restored to service; *Cart. de Berdoues*, no. 459 (1211); cited above chapter 3, n. 78. Former monasteries or hermitages, like Campagnes (Villelongue) and Artigues (Berdoues), were reduced to granges where altars continued to be maintained; see Berman, "Cistercian Development," 202–203.

[5] See plates in Berman, "Fortified Granges," 131, 132, 135, 138–40, and frontispiece here.

[6] Rodez, Soc. des lettres, Bonnecombe, "Cart. d'Iz et Bougaunes," nos. 252 (n.d.), 278 (n.d.), 47 (1181), 55 (1183), 81 (1184), 82 (1184), 83 (1184), etc.

they are not immune from remodelling or destruction, particularly since they are not designated as historical monuments.[7] Most surviving buildings appear to date from the fourteenth century or later, although precise dating of their various structures and the measurement and assessment of the size and storage capacity of such buildings have rarely been attempted.[8] Fortified towers at many Cistercian grange centers in southern France were built during the fourteenth century when the granges became targets for looting mercenaries during the Hundred Years' War. This suggests that they were recognized by that time as the granaries of what had become particularly rich farms.[9]

Little can be said about the maximum or minimum acreage of the southern French granges. Those which were converted in the mid-thirteenth century into military settlements called *bastides*, primarily on the border with the English in Gascony, seem to have included between five hundred and eight hundred *arpents* (425 and 680 acres respectively), as is the case for Gimont's grange of Franqueville and Bonnefont's of Apas, respectively.[10] These examples, however, if anything, may be smaller than the average Cistercian grange, since granges chosen by the monks to be converted into *bastides* were probably among their least productive holdings.[11] Another indication of the relative size of Cistercian granges comes from the very large prices which they brought at the sale of monastic holdings by the Revolutionary government in the late eighteenth century; the amounts suggest that those Cistercian granges which continued to be exploited in the fourteenth and fifteenth centuries were among the largest ecclesiastical estates in many regions.[12]

---

[7] Bonnecombe's grange of Bonnefon was only recently destroyed: Raymond Noël, *Dictionnaire des châteaux de l'Aveyron* (Rodez: Subervie, 1971) 1: 139–41.

[8] See Marie-Anselme Dimier, "Granges, celliers, et bâtiments d'exploitation cisterciens," *Archéologia* 74 (1974): 46–57; Marguerite David-Roy, "Les granges monastiques en France aux XIIe et XIIIe siècles," *Archéologia* 58 (1973): 53–62; Walter Horn and Ernest Born, "The Barn of the Cistercian Grange of Vaulerent (Seine-et-Oise), France," *Festschrift Ulrich Middeldorf*, ed. Antje Kosegarten and Peter Tigler (Berlin: de Gruyter, 1968), 24–31; Gratien LeBlanc, "La grange cistercienne de Fontcalvi (Aude)," *Fédération Historique du Languedoc . . .* (Montpellier) (Sète: Fédération, 1956–57), 43–57, and idem, "La grange de Lassale. Etude historique et archéologique d'une 'grange' cistercienne," *Fédération des sociétés savantes*, Montauban, (Montauban: Fédération, 1954), 3–16.

[9] Berman, "Fortified Granges," 124–46.

[10] Higounet, "Cisterciens et bastides," *Le Moyen Age* 56 (1950): 72–74; on the English-founded equivalents, see Maurice Beresford, *New Towns of the Middle Ages: Town Plantations in England, Wales, and Gascony* (New York: Praeger, 1967), esp. 348–75, which discusses English-founded *bastides* in Gascony; Beresford does not discuss the French foundations.

[11] For example, one small grange converted into a *bastide* was that of Bedored, held by Berdoues, see *Cart. de Berdoues*, nos. 704–46. On the other hand, the granges of Grandselve which became its *bastide* of Grenade were extensive properties; see Higounet, "Cisterciens et Bastides," 267–72 in *Paysages*; on northern French granges, Higounet, "Essai," p. 169.

[12] For the Rouergue, these prices are readily available. Cistercian grange values in that province can be compared to those of other orders. For example, Bonneval's grange of les Gallinières was sold for 411,910 livres, in comparison to the hospital of Aubrac's largest grange of les Bourines, which sold for 382,000 livres at the Revolution, Nöel, *Châteaux*, 1: 434; 2: 104.

Even if they were readily available, actual acreages of Cistercian granges would tell little about their productive capacity in the twelfth and thirteenth centuries, given our limited knowledge of medieval yields and of the technology used or crops planted. Perhaps a more useful way to assess both grange size and the improvements in yields deriving from grange agriculture is to consider the number of peasant farms taken over by the monks—the *mansi* or *casaux* which were consolidated into an average grange within a monastic patrimony. Numbers of peasant farms incorporated (and, by inference, peasant families replaced) do reflect to some extent the productive capacity of the land which such granges comprised. Fortunately, there are documents surviving for the abbey of Bonnecombe in the central Rouergue which include lists of the *mansi* held ca. 1225 by two of its granges—those of St. Félix and Iz.[13] At St. Félix, the monks held total or partial rights in twenty-six *mansi* by that date. At Iz, where land had begun to be acquired somewhat earlier, the monks held total or partial rights in thirty-five *mansi* or their equivalents: that is, in one case the consolidated Cistercian grange would replace twenty-six peasant farms; in the other, it probably eventually replaced at least thirty-five such holdings. Thus, a typical grange belonging to a large abbey by the mid-thirteenth century had incorporated about thirty peasant farms all located within two or three adjacent parishes.[14]

The considerable size of the granges held by Bonnecombe in the early thirteenth century, however, may be somewhat deceptive, for these were the holdings of a prosperous house in the generally prosperous line of Clairvaux.[15] Houses in other filiations may have had smaller granges or fewer such properties, or have gained them at a slower pace. Moreover, in the same way that it is difficult to compare acreages of holdings which varied in soil type and fertility, it is difficult to compare granges or estates on the basis of numbers of peasant tenures from region to region; not only did the names of such holdings vary, but there may have been different ideals as to the size of a typical peasant family to be supported and the

---

[13] Rodez, A. D. Aveyron, 2H Bonnecombe, 39-1, no. 14bis (films 98B and 98C) gives the following information for the holdings at Iz, Oneth, and St. Felix:

Grange of Iz: complete rights in 27 *mansi*, allodial rights in 4 additional *mansi*, complete rights in 1 *caputmansus*, complete rights in 2 *appendariae*, and complete rights in 1 *ausedatz* (?)

Lordship of Oneth: complete rights in 7 *mansi*, 1 *bovarium*, miscellaneous fields.

Grange of St. Felix: complete rights in 16 *mansi*, allodial rights only in 1 *mansus*, complete rights in 1 half-*mansus* and in three parts of five other *mansi*, allodial rights only in three parts of two other *mansi*, *fevum* rights in 1 *mansus*.

[14] For Iz, this included the three parishes of Iz, Limosa, and Ampiac: Rodez, Soc. des lettres, MS Bonnecombe, "Cart. de'Isz et Bougaunes," no. 251 (1194).

[15] Perhaps the political importance of certain monks and abbots of houses like Grandselve and Fontfroide, like Arnald Amalric, leader of the Albigensian Crusaders, abbot of Grandselve, then abbot of Cîteaux, and later archbishop of Narbonne, or the famous Peter of Castelnau from Fontfroide whose assassination set off the Albigensian Crusade, may account to some extent for the eminence of these houses. See Raymond Foreville, "Arnaud Amalric, Archévêque de Narbonne: 1196–1225," *Narbonne: Archéologie et Histoire* (Montpellier: Fédération, 1973), 129–46.

level of subsistence to be obtained, as well as differences in the amount of land on estates still devoted to lord's demesne. The relationship between strictly agricultural production and pastoralism, fishing, and hunting and gathering also undoubtedly varied from the Rouergue to Gascony.[16] The number of *casaux* incorporated into a grange by the Gascon abbey of Berdoues, for instance, was closer to fifteen or twenty as compared to the thirty *mansi* included in Bonnecombe's granges in the Rouergue. This may mean that *casaux* were generally larger than *mansi*, but it is as likely that Berdoues's granges were smaller than Bonnecombe's.[17] Similarly, in acquisitions for its abbey site and home farm, the abbey of Villelongue near Carcassonne probably replaced no more than eight tenures (*tenementa*) in the *villa* of Villelongue and perhaps as many in a neighboring village or parish of approximately the same size. The term *tenementum*, however, is more vague than *casal* or *mansus* and may sometimes have comprised something more than a holding intended to support a single peasant family.[18]

The number of parishes into which Cistercians extended their acquisitions for a single grange might be as high as three or four, as in the examples of Bonnecombe's grange of Iz, near Rodez, which had rights over land owing tithes to churches at Iz, Limosa, and Ampiac, or that abbey's grange of Bernac near Albi, with acquisitions of tithes in the parishes of Gondor, Bernac, Pleus, and Maxières.[19] On the other hand, no more than two parish churches seem to have been actually acquired by the order in the vicinity of any one grange.[20] Unfortunately, there is nothing to prove the absolute size of a parish or indicate the number of peasant households it had included; moreover, it is rarely possible to verify that the monks had acquired all cultivable land within a specific *villa* or parish.[21] It is also possible to

---

[16] The differences, for instance, between areas of Gascon or Basque settlement and those farther east are obvious in any study of the region; see for instance, Natalie Zemon Davis, *The Return of Martin Guerre* (Cambridge, Mass.: Harvard University Press, 1983), 6–24, 51–54, who discusses inheritance pattern variations.

[17] This estimate is made by dividing the total different *casaux* listed in the Berdoues cartulary index by the number of granges described in that volume.

[18] Paris, B. N., Coll. Doat, vol. 70, Villelongue, no. 1 (1164), and nos. 27–33 (1165), and Berman, "Dating," forthcoming; *tenementum* is defined by Niermeyer, *Lexicon*, 1017, as a tenement, a landed estate, or a territory subject to a public authority.

[19] Rodez, Soc. des lettres, MS Bonnecombe, "Cart. d'Iz et Bougaunes," no. 251 (1194), and Rodez, A. D. Aveyron, 2H Bonnecombe, "Cart. de Bernac," no. 106 (1198).

[20] Of the granges acquired by the monks of Bonnecombe, four of them were associated with two churches acquired by that abbey, three with one church acquired, and two with no church acquired. See Berman, "Cistercian Development," 201–203.

[21] See *Cart. de Bonneval*, nos. 52 (1183), 8 (1168), and *Cart. de Silvanès*, nos. 411 (1147), and 231 (1164): "ego Petrus, predicte Rutenensis ecclesie episcopus, . . . dono . . . monasterio Salvaniensi et tibi Poncio, abbati, . . . ecclesiam Sancte Crucis de Serrucio et ecclesiam Sancti Amancii de Senomes, ita videlicet ut nemini liceat sacerdoti prephatas ecclesias sine consensu et consilio abbatis Salvaniensis . . . Laudo . . . donationem ecclesie de Genciaco quam fecit vobis dominus Ademarus, predecessor meus. Verumptamen sacerdotes predictarum ecclesiarum habeant et possideant quicquid ad ipsas ecclesias pertinet vel pertinere debet, hoc excepto quod de laboribus, quos propriis manibus aut sumptibus colitis, sive de nutrimentis vestrorum animalium nullus omnino clericus vel laicus a vobis aliquid accipiat nec exigere presumat," which came after land acquisition in the three parishes.

infer the size of granges from the total number of holdings incorporated or donations involved. For example, the grange of Mercurières belonging to Valmagne was acquired in forty-nine different donations in at least eight different territories, several of them undoubtedly separate parishes.[22] The monastery of Léoncel had acquired five granges in a much smaller number of donations; only twenty-four total recorded conveyances were made to the monks for those five granges by 1176, although many additional conveyances may have been made after that date.[23] The consolidation of each of Gimont's granges required, on the average, 135 separate conveyances of land and even a small house like Calers has preserved sixty charters of land conveyance.[24] Indeed for any specific abbey the charters show that Cistercian granges were not acquired in a single acquisition, and that granges tended to be larger than any individual *villa* or estate which had preceded them.[25]

There is abundant evidence from the cartularies that successful Cistercian houses in the region generally owned a number of such consolidated holdings which they called granges.[26] The number of granges held by any Cistercian house at a specific time is most reliably established from papal confirmations of an abbey's possessions, when such documents are available. In some cases, such confirmations are the only surviving evidence of an abbey's twelfth- and thirteenth-century land holdings. An obvious difficulty in using such confirmations, however, is that they were based on lists of properties provided by the monks for papal confirmation. There is always the possibility that the monks were overstating their case and had "inflated" villages into granges, or single holdings into substantial ones. Nonetheless, where it is possible to compare, charter and cartulary evidence confirm the list of properties in papal confirmations, although it is difficult to assess how complete acquisitions were at any "grange" at the time of confirmation.[27] Papal confirmations of property lists, in addition to listing granges, generally also included a restatement addressed to the individual house of the papal privilege of tithe exemption.

---

[22] Montpellier, A. D. Hérault, "Cart. de Valmagne," vol. I, fols. 39r–52v, vol. II, fols, 154r–155v, at Mercurières, Font Marcel, la Côte Asiné, Mont Ventalos, Pardiniac, Valle Paillède, Gorc Centones, and Las Lobatières.

[23] *Cart. de Léoncel*, no. 25 (1176).

[24] Toulouse, A. D. Haute-Garonne, H Calers, 4–20; the cartulary of Gimont records for the abbey and home farm 141 conveyances, dating from ca. 1142 to 1206, etc.

[25] In the commune in which the abbey of Gimont is found today, there are four churches; see *Cart. de Gimont*, intro., p. xii. Even when complete parishes were not involved, Cistercian cartularies always mention many more parishes than granges; see Berman, "Cistercian Development," Table 1 on 197, for the numbers of churches acquired.

[26] For abbeys with assessments above 15 *livres*, from Appendix I, Table A, numbers of granges can be estimated for six. Bonnecombe had seven, Bonneval nine, Boulbonne nine, Grandselve eleven, Belleperche six, and Valmagne ten. Fontfroide had at least fourteen (but according to Grèzes-Rueff, "Fontfroide," 267, more than twenty).

[27] For example, only a handful of documents from the 1160s and 1170s found in the Valmagne cartulary concern its grange of Burau, but this is considered a grange in the 1185 confirmation of rights there: Montpellier, A.D. Hérault, "Cart. de Valmagne," vol. I, fol. 6r ff., and fols. 83v. ff.

Where a series of such confirmations has been preserved, they can show how quickly a monastic property was acquired. For example, according to papal privileges the abbey of Léoncel, which was founded in 1137, had four granges by 1165: Combacalida, Cognerio, Pallaranges, and Lenthio.[28] Eleven years later, in 1176, a papal confirmation lists an additional grange at Vulpa along with a number of pasture holdings in the Alps.[29] A similar document for the abbey of Silvanès (incorporated by the Cistercians in 1138) lists properties at Gaillac, Grausone, Margnes, Sauveplane, Fontfroide, Sollies, and Rouzet in 1154: seven distinct granges.[30] An additional grange at Promillac and one at Pardinéguas were added to the list by 1162.[31] For the monks of Fontfroide a privilege of protection dated 1162 from Alexander III lists five granges; by the end of the century, a list confirmed by Innocent III contains an additional nine granges.[32] Local studies show that Fontfroide eventually had more than two dozen.[33]

The best information on Bonneval's granges in the Rouergue is in a papal bull of 1162, listing the eight granges of Puszac, Verruca, Masse, Beaucaire, Tegula, La Serre, Empiac, and Biac: that of 1185 adds Pomers, la Roquette, Paulette, Previnquières, Montégut, and Fraissenet.[34] Similarly, a confirmation of properties for Locdieu, which lists five "granges," probably dates to the early 1180s.[35] For these last two abbeys, Bonneval and

---

[28] For example, *Cart. de Léoncel*, no. 13 (1165): "Preterea quascumque possessiones, quecumque bona idem monasterium in presentiarum iuste et canonice possidet, aut in futurum concessione pontificum, largitione regum vel principum, oblatione fidelium seu aliis iustis modis, prestante Domino, poterit adipisci, firma vobis vestrisque successoribus et illibata permaneant; in quibus hec propriis duximus exprimenda vocabulis: locum in quo ipsa abbatia sita est, cum omnibus suis pertinentiis; grangiam de Cumbacalida, cellarium Sancti Juliani, grangiam de Cognerio, grangiam de Pallaranges, grangiam de Lenthio, cum pertinentiis earum. Sane laborum vestrorum, quos propriis manibus aut sumptibus colitis, sive de nutrimentis animalium vestrorum, decimas a vobis nullus presumat exigere."

[29] Ibid., no. 25 (1176): "Preterea . . . locum ipsum in quo prefatum monasterium constructum est, cum omnibus pertinentiis suis, grangiam de Cohongerio, cum omnibus appendiciis suis, grangiam de Palarangis, cum omnibus appendiciis suis, grangiam de Vulpa, cum omnibus appendiciis suis, grangiam de Lenthio, cum omnibus appendiciis suis, grangiam de Cumba calida cum omnibus appendiciis suis, Malum Montem, Charchaleves, Calmen Medianam, Vallem Lutosam et, adjacens nemus, terram Bierini de Chabiol sitam inter duos folliculos, totum Bion, montana de Muson Sane laborum vestrorum . . ."

[30] *Cart. de Silvanès*, no. 2 (1154).

[31] Ibid., no. 1 (1162).

[32] Carcassonne, A. D. Aude, MS 211, "Inv. de Fontfroide," fols. 1v ff.

[33] François Grèzes-Rueff, "L'abbaye de Fontfroide et son domaine foncier au XIIe-XIIIe siècles," *Annales du Midi* 89 (1977): 253–80, maps on 260–67. The maps have a long list of places indicated as granges, among them: Montlaurès, Pradines, Livières, Montredon, Taura, St. Victor de Montveyre, Salavert, Durban, Canemals, Poujouls, Haulterive, Assou, St. Martin, Ste. Eugenie, were early granges, see Carcassonne, A. D. Aude, MS 211, fols, 1B, 4A, and Grèzes-Rueff, p. 267. Although Fontfroide benefited from its earlier Benedictine history and from the incorporation of churches like Montlaurès and Montveyre which probably brought granges with them, perhaps no more than fifteen of the above-listed granges were held by its monks by the middle of the thirteenth century.

[34] *Cart. de Bonneval*, nos. 3 (1162) and 70 (1185); the omission of some of the four granges may reflect name changes.

[35] *Documents concernant l'abbaye de Locdieu*, ed. anon. (Villefranche-de-Rouergue: Société des Amis, 1892), no. 44 (ca. 1181–85).

Locdieu, the papal confirmations provide the most reliable evidence of the extent of acquisitions, since there are few surviving twelfth-century documents concerning those houses.

Elsewhere for the Rouergue, cartulary evidence supplements the papal bulls. For example, one of the wealthiest abbeys in the study, Bonnecombe, had five granges according to a bull of 1178, but one of them, the "grangia de Manso Dei," does not seem to correspond to any property indicated in the extensive surviving archives for Bonnecombe.[36] A papal confirmation of 1220 lists seven granges; four of them are new names, but some of these may be alternate names for the ones mentioned earlier.[37] The confirmation of 1244 lists eight granges, but leaves out Magrin, making nine.[38] Surviving archives also seem to indicate that Bonnecombe had at least eight if not nine large granges by the middle of the thirteenth century.[39] Grandselve, near Toulouse, had an even larger number. The table in a recent article by Mousnier lists twenty-five granges, although at least nine of these are documented only after the mid-thirteenth century.[40] Consultation of papal confirmations and the charters suggests that there were at least eleven

---

[36] *Cart. de Bonnecombe*, no. 64 (1178): "Locum ipsum in quo prescriptum monasterium constructum est cum omnibus pertinentiis suis, grangiam de Valleias cum omnibus pertinentiis suis, grangiam de Monte Calvo cum omnibus pertinentiis suis, grangiam de Magrin cum pertinentiis suis, grangiam de Iz cum pertinentiis suis, grangiam de Manso Dei cum pertinentiis suis, vineas quas habetis in parrochia Sancti Ejecii. Sane laborum vestrorum, quos propriis manibus aut sumptibus colitis sive de nutrimentis vestrorum animalium, nullus omnino a vobis decimas vel primitias exigere presumat."

[37] Ibid., no. 72 (1220): "Locum ipsum in quo prefatum monasterium situm est cum omnibus pertinenciis suis; grangiam de Magrin, grangiam de Capella de l'Arbuselo cum omnibus pertinentiis earumdem; grangiam de Montecalvo, grangias de la Funt et de Isch cum pertinentiis earumdem et grangiam della Calm della Garriga cum pertinentiis suis; possessiones Sancti Hylarii et de Cumbs cum pertinenciis suis, cellarium de Vogonies, cellarium de Egez, vineas de Maurivalle; possessiones de Dotencs, domos de Rodez et de Albia; grangiam de Bernac cum pertinenciis suis, cum pratis, vineis, terris, nemoribus, usagiis et pascuis in bosco et plano, in aquis et molendinis, in viis et semitis et omnibus aliis libertatibus et immunitatibus suis. Sane laborum vestrorum, quos propriis manibus aut sumptibus colitis, de possessionibus habitis ante concilium generale sive de ortis et virgultis et piscationibus vestris vel de nutrimentis animalium vestrorum aut etiam de novalibus nullus a vobis decimas exigere vel extorquere presumat."

[38] Ibid., no. 81 (1244): "locum vestrum in quo prefatum monasterium situm est cum omnibus pertinentiis suis; decimas quas habetis in de Cums de Magrin, de Sancto Ylario, de Novacella, Sancti Martialis et Sancto Felice et de Anglars villis; [de Magrin?], de Iz, de Sancto Felicio, de Bonofonte, de Bernaco, de Montcalm, de Vallellas, de Deusido et de Vogaunes grangias cum decimis et omnibus pertinentiis suis; possessiones, domos, terras, vineas et redditus que habetis apud Casamaurel et Honet ac molendinum da Yssens, decimas, possessiones, terras, vineas, redditus et domos que habetis apud castra Mirabellis et Cassanhas ac Capdenaguet et Amellau villas cum pratis, vineis, terris, nemoribus, usuagiis et pascuis in bosco et plano, in aquis et molendinis, in viis et semitis, et omnibus aliis libertatibus et immunitatibus suis. Sane laborum vestrorum de possessionibus habitis ante concilium generale ac etiam novalium que propriis manibus aut sumptibus colitis, de quibus novalibus aliquis hactenus non percepit, sive de ortis, virgultis, et piscationibus vestris vel de nutrimentis animalium vestrorum nullus vobis decimas exigere vel extorquere presumat."

[39] Judging from the archives, the granges held by Bonnecombe by mid-thirteenth century were: Vareilles, Lafon or Magrin, Iz, Moncan, Bougaunes, St. Félix, Bonfont, Bernac, south of the Tarn at Coms or Pousthoumy, and Bar. See below pp. 70–71.

[40] See Mireille Mousnier, "Les granges de l'abbaye cistercienne de Grandselve, XIIe–XIVe siècle," *Annales du Midi* 95 (1983): 7–27, esp. table on p. 27.

largely consolidated granges in Grandselve's hands by 1250.⁴¹ On the Languedoc coast near Montpellier, Valmagne probably had a dozen granges by ca. 1200: those of Valautre, Vairac, St. Paul, Mercurières, les Ortes, Fondouce, Champdesvignes, Burau (St. Martin de la Garrigue), Canvern, Capriliis, Paunac, and Tresmanses, each of which is represented by a chapter in the cartulary.⁴² In Gascony, Berdoues had at least eight granges by 1250 and Gimont had five by the beginning of the thirteenth century, all designated by cartulary sections.⁴³

While cartularies and papal property confirmations indicate the number of granges held by any one abbey, as indicated in Table 3, it is more often the numbers of surviving charters and to some extent their contents which allow us to assess the relative size of individual granges. Where it is possible to estimate both size and number of granges, these numbers give a fairly good indication of each southern French abbey's relative early wealth within the order in that region. The wealthiest houses were those associated with the filiations of Clairvaux (which had eight houses in southern France) and Cîteaux (which had twelve). Each had five houses whose endowment might be considered large. Four of the fifteen abbeys in the line of Morimond might be considered large to middling, while the eight houses in the line of Pontigny all appear to have been very poor, although one of them, Locdieu, began with a number of granges. There also seems to be a correlation between the richest houses of the region and their easy access or proximity to the region's growing urban centers.⁴⁴ Such impressions of how much property any one house had are confirmed by the fourteenth-century Cistercian tax rolls, extracts from which are included in Appendix I.

Within an abbey's patrimony, the geographical distribution of granges varied. Although most granges tended to be concentrated in the vicinity of an abbey and its home farm, granges might be found in a variety of soil and climate areas. Moreover, most Cistercian houses in southern France had at least one grange located outside the area in which the monastery itself was founded and beyond the order's limiting "one-day's journey" from the abbey.⁴⁵ The most distant granges were probably used as transit

---

⁴¹ On the basis of papal bulls and cartularies one would list Bagnols, Comberouger, Coubirac, Lassale, Salet (Lescout?) Boulhaguet, Calcassac, Nonas, la Terride, Vieilleaigue, Villelongue, St. John (Larra?) and Rieumanent; several of these, however, might be considered actually part of the abbey farm.

⁴² Montpellier, A. D. Hérault (film), "Cart. de Valmagne," passim.

⁴³ *Cart. de Berdoues*, passim, and *Cart. de Gimont*, passim.

⁴⁴ The houses with the highest amounts expended for land acquisition, as revealed in Table 2, tended to be in the vicinity of cities, which may have caused neighboring land values to rise. Houses like Bonnevaux, Valmagne, Fontfroide, Grandselve, and Bonnecombe, however, which remained important throughout their history, all had close ties to regional urban centers such as Lyons, Montpellier, Narbonne, Toulouse, and Rodez, respectively.

⁴⁵ *Statuta*, ed. Canivez, 1, 1134, no. 6, "Monacho cui ex regula claustrum propria debet esse habitatio, licet quidem ad grangias quotiens mittitur ire; sed nequaquam diutius habitare," no. 32, "Grangiae autem diversarum abbatiarum distent inter se ad minus duabus leugis," and no. 43, "Monachus vel conversus non debet minui nisi ad abbatias nostri ordinis, neque ad grangias nisi gravis necessitas incubuerit."

stations for goods, animals, and personnel on their way to the towns where Cistercians sold their produce, but a diversity of granges also provided a balance in an abbey's resources, allowing wine or olive production or winter pasture not otherwise possible in the geographical locale of some Cistercian houses.

In this respect, the distribution of Bonnefont's granges is typical. That abbey had seven granges listed in a privilege of Alexander IV dated 1246.[46] Six of these, although obviously with some variation in features, were located within fifteen kilometers of the abbey's site in the Pyrenean foothills of the Comminges, along the upper Garonne River. The seventh was the grange of Minhac, located just west of Toulouse in Gascony and about seventy kilometers from the abbey of Bonnefont itself. Minhac served as a resting place for Bonnefont's goods and animals *en route* to markets in Toulouse, and perhaps as a winter facility for transhumant animals. Later it was transformed into a *bastide*, one of those fortified towns founded by the Cistercians in concert with the French king or his agents on the eve of the Hundred Years' War.[47]

Bonnecombe's granges in the central and western Rouergue were not quite as widely separated, but they too encompassed a variety of soil types and topography. Their distribution suggests that the establishment of a series of varied granges was intended to allow the monks to produce a variety of cereal types, to spread risk, and to stagger agricultural work through the year. However, the geographical distribution of Bonnecombe's granges was probably also related to that of the patrimony of its earliest donors.[48] The extremely fragmented properties making up Bonnecombe's grange of Magrin or Lafon have already been discussed. That grange was located on the relatively easily worked land found on the hilltops above the Viaur valley just west of the abbey itself.[49] The twin to the grange of Magrin was that of Vareilles, sited nearly opposite the first, to the east of the abbey. It was located in an area south of Rodez generally considered most suitable for rye production. As at Magrin, land consolidation at Vareilles was slow; parts of at least sixty different *mansi* were acquired for it in over 130 contracts.[50] Bonnecombe's grange of Moncan, located south of Magrin and Vareilles, was in a hillier, more pastoral region. Acquisitions there included vines near the Tarn River and concessions of pasture,[51] as

---

[46] *Rec. de Bonnefont*, no. 340 (1246). Locations are given by the editors ibid., intro., 28 and the map, on 34.

[47] Higounet, "Cisterciens et bastides," 269.

[48] See the early gifts by the lords of Panat, Cums, and Sévérac: *Cart. de Bonnecombe*, nos. 251D (ca. 1166), 259A (1179), 260C (1180), 263B (1182), 264B (1183), and 265 (1184).

[49] Rodez, Soc. des lettres, Bonnecombe, "Cart. de Magrin," 15 (1178), 65 (1191), 92 (1201), 91 (1203), 85 (1200), 14 (1178).

[50] *Cart. de Bonnecombe*, nos. 259 A (1179), 269 D (1188), 283 A (1202), 271 G (1191), 278 B (1199), 285 B (1204), etc.

[51] Rodez, Soc. des lettres, Bonnecombe, "Cart. de Moncan," nos. 2 (1171), 14 (1177), 55 (1184), 2 (1171), 49 (1183), 72 (1185), 77 (1185), 78 (1185).

well as rights to both lordship and *pagesia* in *mansi* of the parishes of Auriac and Durenque and in the lordship of Calviac.

Some of Bonnecombe's granges produced wheat, others were located on lands which clearly produced rye, and still others had more mixed cultivation. The abbey's grange of Iz, for example, although it was located in an area known more for wheat production today, also probably produced rye; conveyances even mention a *mansus al Scgalairil*.[52] Iz was sited north of the Aveyron River on the dryer soils of the Causse Comtal, a limestone plateau north of Rodez. Bonnecombe pastured its animals in the vicinity, in addition to cultivating both wheat and rye there, while nearby in the *vallon* of Marcillac, the monks had vineyards at Bougaunes.[53] West of Iz, on the site of the earlier village of St. Félix, Bonnecombe created another cereal-producing grange in the early part of the thirteenth century. Between these two larger granges of Iz and St. Félix, were several smaller holdings belonging to the monks at Ruffepeyre and Puechmaynade.[54] In the Albigeois, Bonnecombe also had granges at Bar near Najac and between Albi and Gaillac at Bernac.[55] Bonnecombe also eventually acquired another grange in the area south of the abbey when the grange of Bonnefon was purchased from Bonneval in the early thirteenth century. That grange was created by Bonneval out of properties in the vicinity of Naucelle in the southwestern Rouergue; there the monks, first of Bonneval and later of Bonnecombe, had extensive pasture rights but also cultivated rye.[56]

Bonneval, which originally held this grange near Naucelle, south of Bonnecombe, is another example of how a Cistercian monastic patrimony might exhibit considerable diversity both in size and location of its granges. Bonneval, located in the northern Rouergue, had granges in two distinct areas

---

[52] On varieties of cereals, see charters where rents are described as payments in kind. For instance, for Magrin and Vareilles rye (*siligo*) is mentioned: Rodez, Soc. des lettres, "Cart. de Magrin," no. 145 (1250); for Iz, eight *sestiers* of wheat (*frumentum*) and five of rye (*siligo*) are mentioned: Rodez: Soc. des lettres, "Cart. d'Iz et Bougaunes," no. 43 (1179); for this and the following, more detail is found in Berman, "Administrative Records," 207–18.

[53] Rodez, Soc. des lettres, Bonnecombe, "Cart. d'Iz et Bougaunes," passim.

[54] Rodez, A. D. Aveyron, 2H Bonnecombe, liasses 39-1 and *Cart. de Bonnecombe*, no. 8 (1221), the conveyance of the church at St. Felix: "M.CC.XX.I., VI. idus julii. Noverint presentes et futuri quod nos P., Dei gratia Ruthenensis episcopus, reducentes ad memoriam religionem et devotionem que, inspirante divina gratia, viget et est vernans modernis temporibus in monasterio Bonecumbe, cum teneamur dictam religionem et devotionem non tantum defendere set etiam confovere et de bono in melius augmentare, de consilio Capituli Ruthenensis ecclesie, donamus et donando concedimus Domino Deo et beate Marie et omnibus sanctis et vobis venerabili fratri B., Dei providencia abbati Bonecumbe, et devoto vestro conventui vestrisque successoribus ecclesiam Sancti Felicis juxta pontem de la Moleda positam cum omnibus juribus et appendiciis compententibus ecclesie memorate, quam vobis damus et concedimus cum omni libertate, salva episcopali reverentia, in perpetuum possidendam; nomine cujus ecclesie dabitis nobis nomine census XV solidos Ruthenensis monete in festo beati Andree annuatim, quos XV solidos annuos vobis placuit nobis dari et etiam obtulistis." See also, Rodez, Soc. des lettres, MS Bonnecombe, "Cart d'Iz et Bougaunes," nos. 145 (1191) and 235 (1231).

[55] Rodez, A. D. Aveyron, 2H Bonnecombe, "Cart. de Bernac," fol. 3, "Memorialium possessionorum de Bernacium," and 2H Bonnecombe, liasse 12, no. 1 (1249), and *passim*.

[56] *Cart. de Bonneval*, nos. 29, 39, etc. are summaries of the charters which were handed to Bonnecombe when the sale was made.

of climate, soil, and topography, even after the sale of the grange of Bonnefon (near Naucelle) to Bonnecombe. Some of Bonneval's granges, such as Biac, Bonauberc, and la Roquette, were on the high Aubrac plateau which adjoins the Auvergne, where winters are long and growing seasons short, but summer pasture abundant. Other granges, like Masse and Puszac near the abbey itself, were on the very southern edge of the Aubrac plateau overlooking the Lot River. The granges north of the Lot River, including the pastoral granges in the central Aubrac, were balanced to some extent by such properties as les Gallinières, la Vayssière, and Séveyrac, all located south of the Lot River on the Causse Comtal north of Rodez (and not far from Bonnecombe's granges of Iz and St. Félix); these granges on the Causse, like that of Bonnefon south of Bonnecombe, produced a variety of cereals on fertile fields in a region having less extreme weather than the Aubrac, and also served as winter stations for animals which could not survive all year round in the cold and snow of the high elevations.[57]

In the southern Rouergue, too, the abbey of Silvanès had diverse granges. One group of its granges (Promillac, Rouzet, and Grauzou) was in the valleys near the abbey site; others (Margnes and Lassouts) were located in the mountains of the Cévennes to the west and south of the abbey where the monks held pasture rights; others, like the grange of Fontfroide, were in the area to the east of the abbey on the route leading up into the pasture lands of the Causse de Larzac. Finally, Silvanès had a grange at Sauveplane on the Languedoc coastal plain which was at least forty kilometers from the abbey. That grange seems to have included both cereal producing lands and pasture, but it was probably also possible to grow olives there. Sauveplane must have served as a way station for goods and animals *en route* to market in Montpellier.[58]

This variation in grange type among the patrimonies of the three major Rouergat abbeys reflects the diversity in Cistercian grange holdings for abbeys from throughout the Midi. In some cases, this diversity is seen as a variation in climate and elevation—very high granges matched with those in the plains; in other cases, there was slighter variation in elevation, but a change in location from one side of a riverbank to the other, less flooded side, made a dramatic difference in the agricultural risk involved. Some granges, like Bonnefont's grange of Minhac, are remarkable because they were located at such a considerable distance from the abbey site (over seventy kilometers), where they served primarily as transit stations for goods and animals on the way to market. Diversity of agricultural production must also have resulted where variation in the elevation of Cis-

---

[57] Bonneval's granges are indicated on Map 4. La Roquette and granges near it were at an elevation of about 800 to 1000 meters, whereas les Gallinières was at only 650 meters above sea level. Surviving sources are all found in either *Cart. de Bonneval*, passim, and especially papal confirmations no. 3 (1162), and no. 70 (1185) or Rodez: A. D. Aveyron, 2H Nouv. ser. 19 "Inventaire de Bonneval," passim.

[58] *Cart. de Silvanès*, intro., pp. xlviii-lv, and Berman, "Silvanès," 301-302.

tercian granges is found, for instance, for the abbey of Villelongue, located just south of the Cévennes, near Carcassonne. Villelongue had a high grange at its original site of Campagnes on the Montagne Noire, but had itself moved southward down into the lush valley of Villelongue in the direction of Carcassonne where much of its agriculture was practiced.[59] Similar effects on production must have resulted for the monks of l'Escaledieu, which had a grange (the original monastic site) very high in the Pyrenees as well as granges on the Gascon plain. The abbey itself had moved to a point midway between the two, still in the highlands, but much more easily accessible than the original site.[60]

Monasteries on the Languedoc coast tended to match their coastal granges with others located in the highlands away from the intense heat of the Mediterranean summers. The monks of Valmagne, for instance, had a series of granges like Mercurières, Valautre, and Vairac, not far from the abbey in the vicinity of Agde. The hot dry climate of the Languedoc coast affected all of these granges; many also had rocky outcroppings or *garrigues* adjacent.[61] Farther north, Valmagne also had a grange in the mountains of the Cévennes at Canvern and one in the forest of Burau, both not far from Silvanès's properties. There seems to have been some rivalry between the two abbeys both in terms of property acquisition and pastoralism there, closer to Silvanès, where Valmagne's monks must have sent its animals to escape the heat of the Languedoc plain.[62] Like Valmagne's granges, most of Fontfroide's granges were quite near the latter abbey, which was located in a particularly bleak area of the hills of the Corbières just south of Narbonne, and in an equally hot region.[63] Perhaps as a result of the unbearable summers, Fontfroide's monks were particularly active in moving animals long distances into the highlands: into the Cévennes, the Pyrenees, and even to Catalonia, where the abbey had daughter-houses if not granges.[64] In Provence, a difference in climate and elevation between a high and isolated abbey site and other property at lower locations owned by the abbey of Léoncel eventually led to the annual winter migration of the monks themselves to the grange or priory of Pardieu, located at a considerably more clement elevation, after its acquisition in 1194.[65]

---

[59] Paris, B. N. Coll. Doat, vol. 70, Villelongue, nos. 1 (1165), 2 (1149), 3 (1149), 3 (1149), 5 (1149), 6 (1149), and 7 (1149); for the correct dating of no. 1, see Berman, "Dating," forthcoming.

[60] LeBlanc, "Repartition," 585–95, and Higounet, "Cisterciens et bastides," *Paysages,* 268–69, map 26 on 270.

[61] Montpellier, A. D. Hérault, film "Cart. de Valmagne," passim.

[62] Ibid., vol. 1, fols. 76r–86v.

[63] Grèzes-Rueff, "Fontfroide," map on 260–61.

[64] Ibid., 257–58, 275–78.

[65] *Cart. de Léoncel,* no. 53 (1194) required the monks to reside in winter at Pardieu. There are other instances from the later Middle Ages of abbeys moving temporarily or permanently to the site of a grange. For example, the abbey of Ardorel moved permanently to its grange of la Rode after the Wars of Religion, and Boulbonne eventually settled at its grange of Tramesaigues. See Lacger, "Ardorel," col. 1620, and Ch. Cathala, "Boulbonne," *Dictionnaire d'histoire et de géographie ecclésiastique* (1938), 10: col. 64.

Although Grandselve, north of Toulouse, had no early granges at a great distance from the abbey site, it did have summer pastures in the mountain valleys of Ravat and rights in the market town of Tarascon-sur-Ariège which were over one hundred kilometers south of the abbey, as well as olive groves near Perpignan at a distance of nearly two hundred kilometers by way of Carcassonne from Toulouse and the abbey.[66] Grandselve, however, is more remarkable for the variation of its granges within a relatively small area. Some were located along the riverbed on the east of the Garonne, an area subject to floods, where the river frequently changed its course. Others were above that river's bluffs on its west bank, where the abbey stood little risk of losing crops to flooding.[67] By having properties in such diverse locations, none of them very far from the abbey, Grandselve's monks, unlike the peasant owners whom they had replaced, could afford some risk of flooding on what were probably otherwise exceptionally fertile soils. The fact that peasants were less able to endure such disasters may explain why such lands came relatively easily into Grandselve's hands.

In addition to protection from the threat of bad weather which might destroy an entire crop in one place, having granges in a variety of areas within a reasonable distance from one another allowed more efficient use of Cistercian labor. If both the coming of spring and the ripening of crops varied slightly from one property to the next, it became possible to stretch out or stagger the periods of most intensive work during the agricultural year, by moving laborers from one grange to another as need arose, without unduly delaying either planting or harvesting. If Cistercian houses were successful in moving personnel about in this fashion, they may have eliminated the seasonal underemployment considered endemic to medieval agriculture,[68] and also have reduced the number of day laborers needed at peak agricultural seasons. The only evidence from early surviving documents that the order's laborers were moved about, however, concerns Grandselve's olive groves in the Roussillon; properties there were overseen by a resident agent during the year who assumed that laborers from Grandselve would assist him at harvest time.[69] If a rotation of laborers occurred elsewhere, it is not documented in the land acquisition records.

This ability to diversity crops and to use labor economically because of an abbey's ownership of a variety of granges, was only one aspect of the efficiencies of the Cistercian grange and its advantages over the agriculture

---

[66] This reference to Grandselve's property is found in a *convenientiae* preserved among Boulbonne's documents, Paris, B. N., Coll. Doat, vol. 83, Boulbonne, no. 99 (1191); on olive groves, see n. 69 below.

[67] According to one charter, rents owed by Grandselve would not be due if there were flooding: Paris, B. N. Latin MS 11011, Grandselve, no. 161 bis (1172).

[68] See Duby, *Early Growth*, 78–88, on the problems of earlier estate management and how poor use of labor resources contributed to the poverty of early medieval agriculture.

[69] See Paris, B. N., Latin MS 9994, Grandselve, fols. 223r–25v, for the concessions of rights at St. Cyprien near Elne. This is a *précis* of many documents perhaps making up another cartulary, now lost. Assistance in the olive harvest by the monks is mentioned ibid., no. 752 (1157).

of the order's neighbors or predecessors in southern France. It is also possible to assert, even without having any sort of production records, that just the introduction of the larger grange size, with its efficiencies of scale and direct management, must have immediately improved net yields and increased savings to the monks. That there was an economic advantage to the grange *per se* is implied by the acquisition policy of the monks as described earlier, which clearly sought to transform smaller farming units into large-scale granges. Assuming all else equal, the smaller size and fragmentation of the earlier units had tended to contribute to diminishing agricultural returns, while the reconstitution of those holdings increased profits.[70] This is true both with regard to fragmentation of land units themselves and to the excessive dispersion of lordship and other rights over that land. Several examples of how larger-scale granges could be more efficient can be mentioned. First, the compact grange had more land actually under cultivation, since it was possible to eliminate boundary ridges, hedge row divisions, and so on. In addition, even without increased acreages, larger fields took less time acre per acre to plow than smaller ones, since start-up and travel times and the number of turns by plows and teams were reduced. Thus, the monks could probably have cultivated the same acreages using less labor, meaning that production costs were decreased—fewer hands working the land meant fewer mouths to feed and higher net yields after labor costs. On large Cistercian farms there was also less fencing needed and fewer chances of animals straying into planted fields since fencing could be more thorough; this meant less loss to accidents.

Obviously, conducting larger-scale agriculture without interference from other owners also contributed to its increased profitability. Direct cultivation on lands where other claimants to the harvest had been eliminated cut down on the interference of middlemen involved in the collection of the harvest or expecting shares of it. Both lord and peasant had lost when crops spoiled in the fields while awaiting tithe or rent collection; with granges totally under Cistercian control it was no longer necessary to await other claimants to the harvest before bringing it under cover. Earlier, lords had often had to pay somone a percentage of rents on more distant holdings in order to collect them at all. Peasants too had had such costs; earlier, they had been expected to invest time and the labor of their beasts in transporting produce to far-off lords. The monks, in contrast, having rid themselves of all such outsiders with claims to their harvests, could collect their crops quickly, and sell what they themselves did not need under privileged conditions (without market taxes and bridge, road, and river tolls) in the nearest markets. Thus, simply by consolidating holdings and

---

[70] The efficiencies of increased scale in agriculture tend to be a commonplace of the literature, but there is considerable analysis of the subject, particularly for the late medieval and early modern period. See, for example, Donald McClosky, "The Economics of Enclosure: A Market Analysis," *European Peasants and their Markets. Essays in Agrarian Economic History*, ed. William N. Parker and Eric L. Jones (Princeton, N.J.: Princeton University Press, 1975), 123–60.

reassembling fragmented land units, the Cistercians in southern France would have produced larger net harvests than the estates and farms which had preceded them, even with the same nominal acreage.

Having land "in hand" also allowed the monks to rationalize both agricultural production (the types of crops grown on any particular grange) and technique (the numbers of plowings, the rotation, manuring, etc.), as well as to reap all the rewards of any capital investment which they made. This freedom to organize production more efficiently extended beyond adjustments to seasonal schedules of plowing and harvest, to a total control over what crops the monks produced. Decisions on planting could be made more rationally, because the monks were responsible to no lord who expected a rent in wheat when the land may actually have been more suited to hay production or to a different cereal.[71] Moreover, monastic production could, to some extent, be geared to market demand—although given its constant internal demand for bread, the order must have continued to produce cereals. That the monks could choose what crops to produce did mean that they could introduce such industrial crops (for which there was increased demand in the twelfth and thirteenth centuries) as dyes, flax, or hemp.[72] In addition, although cereal prices seem to have stagnated in many places in western Europe during this period of increasing wealth (and the accompanying change to a higher protein diet), the order's commercial and tax advantages in selling surplus cereals in local markets probably meant that cereal production continued to be profitable for Cistercians.

As already mentioned, labor could also be used more efficiently by the order on its new granges. In comparison to their predecessors, the monks and *conversi* did not have the constraints on their time that a peasant had had in the form of labor services still owed to a landlord. Moreover, even if those services had been commuted for cash, Cistercians did not have calls on their pocketbooks for such rents, nor were they, like peasants, compelled to continue to produce crops on land which would have been better left to rest. Indeed, it is possible that some of the more marginal lands which were conveyed to Cistercians in southern France were taken out of cereal cultivation altogether, to be used for viticulture and orchards (probably by being rented out to tenants willing to undertake the planting and hand labor), or irrigated to produce hay, or converted into pasturelands

---

[71] Many lands coming into Cistercian hands still had rents in kind on them, for example, *Cart. de Bonnecombe*, no. 251A (1163), quoted above, chapter 2, note 5, and ibid., no. 251B (ca. 1166): "Notum sit quod ego Petrus Ugo et ego Matfres et ego Austorga, mater ejus, nos simul . . . donamus et concedimus in perpetuum Domino Deo et Beate Marie Candelii et Gausberto, abbati, et vobis fratribus ejusdem loci . . . alodium mansi de Berengairenc et mansi de la Valle et quicquid ibi aliud habemus vel habere debemus, . . . et notificamus censum annualem uniuscujusque mansi suprascripti, vidilicet: I porcum et I multonem et IIII, sestarios siliginis de taverna et II sestarios avene et II panes et I espathlam."

[72] On such industrial crops, see Philippe Wolff, *Commerces et marchands de Toulouse, vers 1350-vers 1450* (Paris: Plon, 1954), 247–52.

until their fertility was restored. The documents available, however, are rarely explicit.[73]

Like the advantages inherent in larger units of land, both efficiencies of scale and the advantages of such total control over the management of agriculture probably automatically increased gross yields at least slightly on Cistercian granges, as well as reducing labor inefficiencies and cutting down on payments to middlemen. Unfortunately, the savings in time and other agricultural efficiencies which resulted from the rationalized, large-scale agriculture instituted by Cistercians on their granges are virtually impossible to document. It is likely, however, that a number of such small improvements may have added up to substantially increased yields. These efficiencies of the grange *per se*, of its size and rationalization, would also tend to magnify the effects of other improvements in Cistercian grange production such as those in agricultural technology. Moreover, some of the advantages of consolidation by Cistercians may have been shared by their neighbors; given the fragmented nature of holdings in areas where Cistercians had acquired land, for it is likely that a concession of one holding to the monks may have allowed a peasant or lord to consolidate land held elsewhere.

Probably the most important advantage of the grange *per se* was the order's ability to take land back under its own management, allowing it to introduce direct cultivation using its own laborers. Indirect proof that the Cistercians took land "in hand" can be found in the land acquisition records. The best evidence that such new management was instituted is found in charters recording concessions of tithes on lands under Cistercian direct cultivation. Concessions of limited rights to tithes "on the products of Cistercian labor or direct management" at specific places were often worded in the terms of the papal and episcopal tithe privileges already discussed; such concessions imply not only direct cultivation by the monks on particular holdings owned by Cistercian abbeys, but give dates by which such cultivation had begun. These documents show, for instance, that Gimont, founded in the 1140s, already had direct cultivation at two different granges by the 1160s, and that Bonnecombe, founded in the 1160s, had its first

---

[73] One contract, a conveyance of pasture to Silvanès, allowed the monks to pasture their animals on inhabited lands within the parish, if given permission by the peasants there, but this refers to rights to pasture on fallow, rather than the conversion of agricultural land to pasture land: see *Cart. de Silvanès*, no. 448 (1164): "M.C.LX.IIII., ego Raimundus Guillermi de Fabrezano et ego Englesa, soror ejus, et maritus meus Petrus de Podio Laurencii, . . . donamus et laudamus et titulo donationis cum hac carta tradimus monasterio Beate Marie de Salvanesc et tibi Poncio, abbati, . . . totum quod habemus et habere debemus vel aliqua persona per nos vel de nobis habet in toto territorio de Las Solz, videlicet terras cultas et incultas, nemora, prata, pascuas, decursus aquarum et recursus et si qua sunt alia ad ipsum territorium pertinencia. Insuper laudamus vobis pascuas in omni terra nostra ad alenda et nutrienda animalia vestra cujuscumque manieri sint, exceptis mansis vestitis in quibus pascere non debetis absque pagensium eorum consensu qui terram colunt."

concession of such rights to tithes on its direct cultivation by 1172.[74] Grandselve may have instituted direct cultivation as early as 1155 and had expanded its own management to half a dozen other sites by the 1190s.[75] Silvanès, in particular, seems to have taken land "in hand" at several granges from a very early date: at Sauveplane in 1147 and at Margnes in that same year.[76] It is reasonable to suppose that at granges even closer to that abbey, the practice of direct management was instituted even earlier, perhaps soon after Silvanès's incorporation into the Cistercian order in 1138.

Such evidence of tithe concessions suggests that at certain sites before the end of the twelfth century, the process of Cistercian consolidation had been carried to its desired conclusion: the institution of the new, large-scale, directly managed agriculture. Direct management was probably instituted in many other places where local tithe concessions were not made in such explicit terms or where documentation is lacking. Indeed, evidence of the presence of *conversi* also implies that land was being taken "in hand." Such evidence of the recruiting of peasants as *conversi* laborers is found in the many charters granting them rights to enter Cistercian houses. By counting charters promising admission as *conversi* for the abbeys of Berdoues, Bonnecombe, Grandselve, Gimont, and Silvanès, it is possible to establish a minimum number of *conversi* for these abbeys for which particularly good records survive. On the average, each of those houses admitted at least eleven *conversi* in the period 1140–1159, thirty-five in the period 1160–1179, and thirty-nine between 1180 and 1199. This means that after 1160, *conversi* laborers were recruited by these abbeys at the rate of about one every twelve months.[77] There would be no need for this recruitment without the direct management of agriculture on Cistercian estates which the presence of such *conversi* implies.

The efficiencies and savings related to larger-scale agriculture and direct management of Cistercian granges were closely tied to a second advantage of grange agriculture as it was practiced in southern France. This was the reduction of agricultural labor costs through the use of lay-brothers or *conversi*, who did the major part of the agricultural work on the granges in place of tenant farmers. The recruiting of these *conversi* and the relationship between their recruitment and land acquisition have already been discussed. The advantages provided by *conversi* to Cistercian agriculture tend to be a commonplace of the historical literature. It is often mentioned that *conversi* laborers were free from the constraints of peasant mentality, from dependence on landlord's demands, from the lack of cash which had

---

[74] *Cart. de Gimont*, VI, nos. 15 (n.d.), 118 (n.d.), 6 (1160), 48 (1160), etc. Rodez, Soc. des lettres, Bonnecombe, "Cart. de Magrin," nos. 57 (1182), 51 (1200), "Cart. d'Iz et Bougaunes," nos. 36 (1172), 126 (1179), etc.

[75] Paris, B. N., Latin MS 9994, Grandselve, fols. 181r (1155), 181v (1155), 114r (1163), 197r (1164), 200r (1170), 205r (1177), etc.

[76] *Cart. de Silvanès*, nos. 390 (1146), 391 (1146), 392 (1147), and 411 (1147).

[77] Only explicit references to admission as *conversi* were counted.

kept earlier peasants from improving land cultivation, and so on; less often is it mentioned that the use of *conversi* lowered dependency costs for labor.[78] What is rarely considered is how much less the dependency costs of *conversi* could be.

The primary reason that the cost of *conversi* labor per hour was less than for the dependent tenants who had preceded them, was that *conversi* were not feeding wives, children, or the elderly, but (in terms of labor costs) had only to be fed and clothed themselves. Because they represented fewer labor-related mouths to feed, the use of *conversi* by Cistercians had the immediate effect of decreasing labor costs and thereby increasing the order's net yields. Leaving aside any increase in gross yields from other efficiencies, the amount of grain left over (after setting aside seed) was automatically higher with *conversi* because they consumed less than peasant families. In other words, the unit cost of subsistence per laborer was significantly lower on Cistercian granges using *conversi* than it had been on the peasant farms which had preceded those granges, even if man-hours expended, efficiency, and yields were exactly the same. This was a major reason for improvement in net agricultural yields under Cistercian management.

To a certain extent the same applies to hired laborers. Assuming that laborers could be hired on short notice at an hourly wage and that an abbey was freed of any responsibility for those laborers when they were not employed by that house (that is, that they were truly seasonal and not dependents), there would be an increase in net yields to Cistercian agriculture because hired laborers had to be paid (or fed) only at peak times of year. They did not have to be supported during periods when they were not needed and there would be a tendency to employ specialists—specialized agricultural laborers, roofers, carpenters, stoneworkers, and so on—rather than using the same casual laborers for different jobs as the seasons progressed and as agricultural tasks increased or decreased. It has already been mentioned that there was competition within the agricultural world in recruiting both hired workers and *conversi*, and that *conversi*, once recruited, although providing a permanent labor force for a lifetime, did not replace themselves. This problem of replacement did not, however, immediately arise, and may not have been apparent at the outset to Cistercian managers.

The lowered dependency cost of *conversi* labor on Cistercian granges is best demonstrated by a hypothetical example. This example assumes that previous peasants had operated at a subsistence or near-subsistence level—not producing noticeable profits for themselves—and that in a normal year on a farm under peasant cultivation the gross harvest would have been expanded wholly among seed, rent, tithes, and the support of the peasant and his dependents. Thus, where labor cost equals the amount necessary to feed a peasant and his dependents on that acreage:

---

[78] Roehl, "Plan," 93. Graves, "Economic Activities," does not consider this.

GROSS HARVEST = SEED + LABOR COST

+ RENT + TITHE + PEASANT'S PROFIT.

Since it can be assumed that in a normal year the peasant lived at subsistence level and that he merely broke even, with nothing to spare or save, this equation can be simplified by excluding PEASANT'S PROFIT to:

GROSS HARVEST = SEED + LABOR COST + RENT + TITHE.

Rearranging this equation, the actual return (RENT) which would come to "owners" of the land, can be expressed as:

RENT = GROSS HARVEST − SEED − LABOR COST − TITHE.

and LABOR COST can be expressed thus:

LABOR COST = GROSS HARVEST − SEED − RENT − TITHE.

If numbers are substituted and a static system assumed, in which lords did not allow their peasants to starve and thus rents were in some proportion to yields, the amount of food necessary to support laborers on the peasant farm can be estimated. Given an acreage of indifferent size, but which produced one hundred bushels annually, with the peasant merely breaking even, a yield/seed ratio of 3:1 (a little higher than usual at this time, but other possibilities are given in Appendix 2), with RENT as the quarter of gross harvest which seems standard in this region, and the ecclesiastical TITHE and its appurtenances equal to only 10 percent of the gross harvest (quite often they were higher),[79] the following numbers can be inserted into the equation:

LABOR COST = GROSS HARVEST (100 BU) − SEED (33 BU)

− RENT (25 BU) − TITHE (10 BU)   or   LABOR COST = 32 BU.

Thus, in this case, thirty-two bushels or 32 percent of the gross harvest is the amount needed each year to feed the laborer and his family, after rents were paid and seed set aside.

In comparison, Cistercian *conversi* without dependents to feed, cultivating the same piece of land with the same yields, would have obviously consumed less food. What was left over could go into the RENT category. If Cistercian laborers only consumed twenty-two bushels instead of thirty-

---

[79] On yields, see Slicher van Bath, *Agrarian History*, Table 2, pp. 328–29, and Duby, *Rural Economy*, 99–101. On tithes, see references below. On rents, the *quartum* is often mentioned in southern French documents; see Elisabeth Magnou-Nortier, "La terre, la rente, et le pouvoir dans les pays de Languedoc," *Francia* 10 (1982): 33–35, where she describes the traditional payment for a *mansus* in the region of Narbonne, Beziers, the Rouergue, and Pays d'Aude as either a tenth on the lands and a quarter on the vines, or a quarter on the lands as well as on the vines, plus a fixed annual cens, plus a payment for use of water, wood, etc., plus an element representing labor services. The *quartum* is also found for the region of Toulouse, see Ourliac, "Réflexions sur le servage Languedoçien," *Comptes rendus de l'Académie* 98 (1971): 585–91; and for olive trees in the diocese of Elne, Paris, B. N. Latin MS 9994, Grandselve, fols. 223r ff.; the rent of a quarter is also discussed in Duby, *Rural Economy*, 217–18.

two (that is, 20 percent of the gross annual harvest on the hypothetical acreage) or only about two/thirds of what the peasant family ate, the ten bushel addition to RENT would mean a 40 percent improvement in net yields after SEED and LABOR COST and TITHE are paid. Thus, the owner's net yield or RENT would be:

GROSS HARVEST (100 BU) − SEED (33 BU)

− LABOR COST (22 BU) − TITHE (10 BU)

= RENT of 35 BU as opposed to the earlier RENT of 25 BU.

If Cistercian laborers consumed only half that which the peasant family did (that is, sixteen bushels), a 64 percent increase in net yield or RENT would have been felt, and so on. What is noticeable here is the very large percentage gain in RENT which occurs as labor cost decreases. This is a striking improvement. (The yields suggested here were not unusual in the Middle Ages, but see Appendix II for examples using different assumptions about yields, rents, and so on.)

This dependency savings for *conversi* labor was a major advantage of Cistercian agriculture and must be recognized as one of the reasons why direct cultivation by lay brothers and hired laborers was so attractive. It was probably especially significant in aiding the rapid expansion of the order's agriculture in its early years, when *conversi* were plentiful and monks worked in the fields and neither group included large numbers of aged and infirm dependents. Even if such dependency cost advantages gradually disappeared as *conversi* aged and monks spent less time doing agricultural work, the initial savings would have been sufficient in many cases to transform a moderate-sized abbey into a large and wealthy one.[80] Whatever difficulties eventually developed in *conversi* recruitment and dependency, lower initial dependency costs for *conversi* laborers were important in allowing a very profitable start to the earliest Cistercian grange agriculture.

An associated and additional advantage of *conversi* may have been that they were more productive man-year for man-year, because these lay brothers, who could be moved from grange to grange and field to field, were not only less seasonally underemployed but more fit for agricultural work. The economic security of life in a Cistercian abbey meant that *conversi*, unlike peasants who were often on the verge of starvation, were healthier and consistently better fed. That probably made them not only more efficient, more innovative, and less lethargic, but more productive.[81] *Conversi*

---

[80] In this calculation, it is only possible to compare Cistercian ownership to earlier lordship, if the monks are treated as landlords living on RENT just as Cluniac ones would have been. Therefore, any labor which they contribute to the production of the fields is negligible or cost-free, in terms of these calculations of LABOR COST.

[81] Duby speaks of such malnutrition in the feudal age in *Early Growth*, 158–83, 183. Recent work by economists and anthropologists on the relationship between malnutrition and low productivity can be found among the papers of such groups as the International Food Policy Research Institute in Washington, D.C., whose recent conferences draw on such earlier studies as H. A. Kraut and E. A. Mueller, "Calorie Intake and Industrial Output," *Science* 104 (1946): 495–97.

were also more efficient because they had better tools and draft animals than peasants could afford, and were not distracted by the cares and concerns of the laity. Finally, at least in the early years, *conversi* must have been highly motivated, seeing their work in the fields as a form of prayer. Thus, the use of these laborers was more than simply a religious ideal, for it contributed markedly to Cistercian economic success, because *conversi* were less costly to support and more efficient and motivated.

A third major advantage of Cistercian agriculture in southern France over that which had preceded it, was that the Cistercians paid no rents, taxes, or ecclesiastical tithes on their production, and paid few tolls and taxes when they moved their surpluses to local markets for sale. Advantageous market and transport conditions were the result of exemptions from market dues and from tolls on the transfer of goods and animals. Such concessions were often granted to them by the most important lords of the region. Valmagne's monks, for example, could take produce without toll to whatever markets they chose between Narbonne and the Rhône. Among the notables making such concessions of toll exemption to that abbey were Raymond, count of Barcelona and prince of Aragon, who gave Valmagne passage rights on the Rhône and in his lands in Languedoc without passage, portage, or usage fees; Raymond Trencavel, viscount of Béziers, who gave the monks rights of passage "by land and sea" without toll, for whatever Valmagne bought or sold; Raymond of Toulouse, whose confirmation to Valmagne of such rights in Provence and Languedoc mentions exemption from tolls on land and water transport as well as from taxes on the abbey's fishing; Alphonse, king of Aragon, whose exemption from tolls for Valmagne dated 1181 mentions transport of cereal and wood on land and water and included specific rights for Valmagne to transport as much wood as was needed to the monastery of Sauveréal on the Rhône River.[82]

In the late twelfth century, the monks of Grandselve also began to acquire rights to pass with their goods down the Garonne River without toll. Grandselve's cartularies include an entire volume of thirteenth-century acts granting and confirming exemptions from tolls on that river from the vicinity of the abbey near Toulouse all the way to Bordeaux. Although seemingly granted at first for the boatload of salt which the Angevin kings had granted to the monks from their salt-works in Bordeaux, the abbey's exemptions expanded to include free transport of wheat, wine, and other goods as well as salt on that river.[83] In addition, Grandselve had rights to move its animals and wool by overland routes and on other rivers to Agen,

---

[82] Montpellier, A. D. Hérault, film, "Cart. de Valmagne," vol. 1, fols. 137v (1148), 151r (1162), 151v (1155), 151v (1181), and 152r (1175).

[83] Paris, B. N., Latin MS 11010, Grandselve, passim, esp. nos. 50 (1199), 62 (1218); and A. D. Haute-Garonne 2H 108 Grandselve, nos. 90 (1190), 131 (1205); Paris, B. N., Coll. Doat, vol. 77, Grandselve, nos. 130 (1247) and 137 (1248).

Condom, and Cahors from the vicinity of Toulouse.[84] Other abbeys having similar rights in their own neighborhoods are discussed in chapter 6.

In addition to having such exemptions from market taxes and passage tolls, most Cistercian abbeys managed to eliminate rents, fees, and tithes owed on land making up their granges by repurchasing such claims over lands making up their granges, part of the process described in chapter 3. As a result, Cistercian agriculture paid no part of its harvest to a lord, incurred no payments to middlemen, and could reap 100 percent of the profit of any investment in agriculture. Under the earlier regime there was less incentive for either peasant or lord to invest in agriculture, whether in new agricultural equipment or better draft animals, since increases in gross yields were shared among so many claimants to parts of the harvest. An investment by the peasant, particularly for good draft animals for instance, would have been lost at his death when his heirs were required to give the best animal to their lord and in some places the second-best to the priest.[85] Similarly, a lay lord, even if he were able to strike new agreements with his tenants which allowed him most of the profits of his investment (and this was not easily done, given established custom), still often paid a percentage of the gross yields from his lands to others: for castle-guard, or *vicaria*, or tithes. He might therefore hesitate to invest capital in agriculture even if he had it available.[86] In contrast, the Cistercian grange had no such disincentives to capital investment, once the harvest was wholly reserved to the new grange owners. Such exemption from taxes, tithes, and other levies on their land, thus provided a considerable incentive for Cistercian investment in agriculture. Moreover, the Cistercians' asceticism and stress on simplicity in religious practice promoted the savings which allowed capital accumulation for such investment.

Tax savings were also substantial in themselves, when considered in terms of net yields, particularly since such levies seem to have been invariably collected on the harvest while it was still in the fields, that is, on the gross harvest. Of all such tax savings, it was the tithe exemption, granted to the order by the papacy in the 1130s, which was of greatest advantage to Cistercian agriculture. Its implementation by the repurchase of existing tithes from previous owners has already been discussed, but its implications are more dramatic than may be immediately apparent. The tithe privilege gave the order an additional tenth or more of the gross harvest since the exemption from tithes also included the first fruits—an additional levy which often increased the tithe to 11 or 12 percent. Like the improvement

---

[84] Toulouse, A. D. Haute-Garonne, 2H 108 Grandselve, no. 120 (1225).

[85] In southern France this payment appears to have been most often in money, but it might amount to the same thing. For the most recent discussion, see Magnou-Nortier, "La terre," 10 (1982):50–55.

[86] Such obscure claims, for instance, are described in *Cart. de Silvanès*, nos. 92 (1165), 107 (1167), 482 (1177), and 375 (1169), etc.

in yields from the introduction of *conversi* laborers, which decreased dependency costs, the exemption from tithes on Cistercian granges improved profits because net yields would increase even if agricultural efficiency stayed exactly the same. The tithe exemption, moreover, must have spurred Cistercians to convert to direct management, since they could claim exemption from tithes only on their stock-raising, and on those lands under the order's own management, or those which it worked with its own labor. As discussed earlier, however, exemption in fact became one more aspect of the land acquisition process.

It is difficult to determine exactly what percentage of the harvest the "tithes" represented, or what produce was subject to them, or exactly how and by whom tithe collection was undertaken, but it is clear that the tithe was a fixed percentage of the gross harvest as it stood in the fields; in fact, it appears that cultivators could not remove their sheaves until the tithe-owner had taken his "tenth."[87] Because it was assessed on gross yields, such an exemption from a 10 (or 11 or 12) percent tax actually represented significantly more in terms of net yields—the yield after seed had been taken out and laborers fed, particularly if yield/seed ratios were low. How large such increases in net yields could be to Cistercians with tithe-exempt agriculture can be shown using the same hypothetical example as above, that of a holding of indeterminate size yielding 100 bushels per year. Without the tithe exemption, the equation for the amount of RENT could be expressed as:

RENT = GROSS HARVEST (100 BU) − SEED (33 BU)

− LABOR COST (32 BU) − TITHE (10)   or   RENT = 25 BU.

Tithe-exempt land, however, gave its owners the 10 bushels attributable to TITHE in addition to the original 25 bushels for RENT. Therefore, with tithe exemption RENT was:

GROSS HARVEST (100 BU) − SEED (33 BU)

− LABOR COST (32 BU)   or   RENT = 35 BU.

This change from 25 bu. to 35 bu. for RENT with tithe-exempt agriculture is a *forty* percent increase in the owner's net yield. If *conversi* labor advantages of perhaps an additional ten bushels are added, the owner's net yield or RENT increases still further, since RENT then equals:

GROSS HARVEST (100 BU) − SEED (33 BU)

− LABOR COST (22 BU)   or   RENT = 45 BU.

That is, for a tithe-exempt farm, with the cost of food for laborers about two/thirds of what it had been under peasant cultivation, the net harvest or RENT to the owner would increase by 20 bushels over the original

---

[87] Le Roy Ladurie, *Peasants*, 80, and Constable, *Monastic Tithes*, 9–31.

owner's *quartum* of 25 bushels—an 80 percent overall gain in net yields. Again, other examples based on slightly different assumptions are calculated in Appendix 2.

Even if the effects of the tithe exemption were not actually quite as dramatic as in this hypothetical example, such gains may often have transformed what had been marginally profitable estates for previous owners into money-making operations for the Cistercians. In almost any situation, no matter what yield/seed ratios are assumed and whether or not yields themselves improved, even if tithes and other rights had had to be repurchased, the combined effect of decreased dependency costs for *conversi* labor and tithe exemption must have caused a dramatic increase in net yields to Cistercian agriculture once land was in hand. These increased net yields made the institution of grange agriculture by the new monks very worthwhile. Moreover, as mentioned earlier, the tithe exemption, like freedom from other taxes and tolls, represented a substantial incentive to the introduction of improvements to Cistercian agriculture. Tithe exemption magnified the profitability of small changes and improvements to yields (the advantages of efficiencies of size, capital investment in tools, new crops or cropping techniques, increased ratios of animals, and rationalized organization) which otherwise may have been marginal.

Moreover, although it is true that the order often had to purchase tithes from previous owners, an abbey could usually acquire them for one-time payments or even for non-monetary considerations from the laity. Even in its acquisition of tithes from other churchmen, the Cistercians usually benefited in the long run. They rarely assigned a share of future harvests to a former owner, instead paying either a one-time sum, or transferring land, or promising a fixed annual rent. Even if such purchase prices or annual payments were not nominal at the outset, they were amortized or reduced over time, as prices rose and money became less valuable through inflation and debasement.[88] Under these circumstances, it seems likely that the large profits to Cistercians from tithe exemptions and apparent losses to the secular clergy (who had undoubtedly expected to benefit more from Gregorian assertions that tithes should be returned to the Church) may explain the vehemence of complaints about Cistercian tithe exemption.[89] It is not surprising that churchmen who had conceded tithes to the order for very little became increasingly irritated that Cistercians, "who had been granted exemption from tithes because they were poor, continued to profit from [that exemption] when they had become rich."[90]

The profitability of exemption from tithes explains such complaints, as

---

[88] On problems of debasement in this region during the period, see Thomas N. Bisson, *Conservation of Coinage: Monetary Exploitation and its Restraint in France, Catalonia, and Aragon (c. A.D. 1000-c. 1225)* (Oxford: Clarendon Press, 1979), 45–119.

[89] Constable, *Monastic Tithes*, 270–306; none of the complaints cited concern southern French houses specifically.

[90] Donnelly, *Decline*, 47, cites this complaint by the archbishop of Canterbury (1171–84).

well as why twelfth-century Cistercian houses carefully sought papal confirmations of their properties which included confirmation of the tithe privilege. The profits involved may also indicate why in the 1180s the General Chapter, citing complaints about the exemption, decided that abbeys should only acquire tithes if they paid tithe-owners for their rights, a position which surely aimed to insure the continuance of the tithe exemption.[91] Interestingly enough it was probably also the profitability of continued tithe exemption which best explains why Cistercian narratives insisted on, and even tended to exaggerate, the "frontier" conditions of their sites—the isolation of their abbeys far from other human habitation.[92] The tithe exemption had probably originally been based on the papacy's notion that it hurt no one (was it believed that the Cistercians founded their houses in wildernesses where tithes had never been assessed?). As it turned out, although appearances are otherwise, efforts to avert complaints and maintain the tithe exemption were generally successful. The major part of the order's land acquisitions had been completed by 1215 when exemption was limited to either authentically "noval" lands or those lands already "in-hand" at that date.[93] Complaints about the tithe exemption by the order's twelfth-century critics, however, do reveal the importance of the tithe exemption to the growth of Cistercian wealth, and how successful the order's economic practices had become.

Several other advantages of Cistercian agriculture over that which had preceded it can be posited, although it is difficult to document the implementation of such possible improvements. Cistercians could improve their agriculture by investment in it, and certainly had tremendous incentives in terms of increased net yields for investing in better tools, drainage ditches, sturdier plow teams, barns, and other grange buildings. The availability of cash for such investment is most readily apparent in land and pasture

---

[91] *Statuta*, ed. Canivez, 1180, no. 1: "Cum divina pariter et humana verba nos admoneant, ut studeamus cupiditatibus nostris imponere frenum, et modum acquisitionibus nostris propter varios status monasteriorum, adhuc veremur necessitatem fratribus nostris imponere, optantes bonum eorum voluntarium esse, et diuturna deliberatione super hoc credimus opus fore. Interim autem propter scandalum gravius, quod super retentione decimarum undique crescit in dies, providemus et firmiter praecipimus, ut quicumque ex vobis ab hoc die et deinceps acquisierit agros vel vineas, ex quibus ecclesiae vel monasteria, seu personae quaelibet ecclesiasticae percipere hactenus decimas consueverant, sine contradictione exsolvant, nisi forte iam donationem exinde receperunt, vel compositionem fecerunt, aut in posterum in pace acquirere poterunt."

[92] Cistercian houses were certainly not above advertising their early poverty or eremitical roots; see *Cart. de Bonneval*, introduction, pp. xv-xix, which describes various foundation legends broadcast by the monks of Bonneval. This would also seem to be the case for Silvanès, which when faced with a threat from the foundation of Candeil to the west in 1152, the incorporation of Valmagne to the south in 1155, and the foundation of Bonnecombe to the north in about 1166, decided to set down an account of its foundation, which is the "Chronicle of Silvanès," discussed below pp. 127-28.

[93] Constable, *Monastic Tithes*, 306. Donnelly discusses the further negotiations on what comprised "noval" tithes: Donnelly, *Decline*, 50 ff.

acquisitions by the order, but there must also have been cash to purchase animals for stock-raising, and smaller tools, and to make other improvements.

One clear example of Cistercian outlay for capital improvements was in the construction and refurbishing of water mills. A Cistercian abbey generally had at least one milling complex at the majority of its granges, and sometimes several additional mills were also associated with such a property. Mills at granges seem to have been intended, in the minds of the order's founders, for internal power needs: grinding flour, fulling cloth, sawing wood, cutting stone, and running bellows and trip-hammers for forges. In addition, however, although several early documents from southern France concern donors who conveyed rights to mills but forbade the monks to constrain peasants from neighboring villages to use those new mills, revenues from mills used by neighboring peasants must often have been an early source of income for the order in southern France; this was certainly true elsewhere, as studies of even the earliest Cistercian houses in regions like Burgundy have shown.[94] Indeed, rights over *moltura* (the flour tax which lords exacted for milling) were sometimes specifically conveyed to southern French Cistercian abbeys along with other lordship over a mill, or with rights to construct new mills or restore dilapidated ones.[95] Cistercians could repair and operate water mills because they had access to timber, craftsmen, tools, and the capital necessary for making expensive repairs. Given the large expenditures involved in mill construction and upkeep, it is likely that the concession of some of these mills to the monks occurred because earlier owners could not afford their maintenance.

Generally construction of a water mill also required considerable capital outlay in amassing rights to the site and to the banks of rivers and areas that would be flooded by the mill pond, as well as in acquiring materials

---

[94] Rights to build mills at St. Jory were conceded by the count of Toulouse who forbade the monks from constraining his peasants to use them: Paris, B. N., Coll. Doat, vol. 76, Grandselve, no. 72 (1185); similarly, *Cart. de Silvanès*, no. 210 (1159), forbade local peasants being forced to use the mill at Promillac. On the use of mills for revenue elsewhere, see Duby, *Saint Bernard*, 107, or *Recueil des pancartes de l'abbaye de la Ferté-sur-Grosne: 1113-1178*, ed. Georges Duby, (Aix-Marseille: Faculté des lettres, 1953), intro., p. 19, where he says, "D'ailleurs, pour satisfaire leurs besoins en céréales, ils disposaient non seulement des blés recoltés sur les vieux essarts, et sur les champs des lisières, mais encore, et peut-être surtout, du revenu des moulins" and see next note.

[95] For example, Montpellier, A. D. Hérault, film "Cart de Valmagne," vol. 1, fol. 150r, two documents dated 1151 are conveyances of *moltura*, although such revenues were forbidden in the statutes; see *Statuta*, ed. Canivez, vol. 1 (1134), no. 9. Claims for *moltura* for other previously existing mills were transferred to Bonnecombe: Rodez, Soc. des lettres, Bonnecombe, "Cart. de Magrin," no. 127 (n.d.). Gimont and Grandselve established a *medium vestum* contract to restore old mills and mill dams at Pinels where Gimont would own half of the existing complex and Grandselve the other; Grandselve could make additional mills there for itself: *Cart. de Gimont*, V, no. 69 (1163): "De Willelmo, qui vocatur abbas de Lombers: . . . Predictus . . . vendidit Bernardo abbati medietatem del mulnars de Pinels cum suis pertinenciis, et cum ingressu et egressu ejusdem pro LXX sol. Morl. tali conventione ut predicte habitores Gemundi et predictus Willelmus restituant per medium vetus [vestum] molendinum et paxeriam."

for construction and maintenance.⁹⁶ Mills constructed or acquired varied considerably in size and power, since rivers harnessed for their water power in southern France varied from the swift-flowing rivulets coming down from the Cévennes to feed mills operated by Silvanès, to the wide and slow-moving Garonne operating floating mills belonging to Grandselve. Technology for mills varied in this region; documents suggest that both overshot and undershot mills seem to have been built or acquired by the order.⁹⁷ In many cases, an earlier mill or mills had stood at the sites where Cistercians would eventually own mills, but it is also clear that capital had to be invested to refit them.⁹⁸

Charters concerning rights to construct mills, or rights to rehabilitate mills fallen into disrepair, show that Cistercian houses in southern France acquired well over one hundred sites for mills, but the number of mills at any site is difficult to gauge.⁹⁹ A document for Silvanès outlines specific

---

⁹⁶ Such donations included vague gifts like "ritbatge ad opus molendini," Paris, A. N. L1009 bis, Grandselve, no. 29 (1165), and rights to take millstones, Toulouse, A. D. Haute-Garonne, Grandselve, no. 273 (1179), as well as specific plans for complex milling projects, *Cart. de Silvanès*, no. 210 (1159), discussed in n. 100 below. Land was given for Berdoues to rebuild a mill at Morlencs: *Cart. de Berdoues*, no. 97 (1167), and that abbey also received three parts of an existing mill at la Bara, ibid., no. 129 (1204), millstones at la Riusas for its grinding needs, ibid., no. 63 (1215), and land at the *caput paxeria* of the mill of Artigues, ibid., no. 224 (n.d.).

⁹⁷ Further description of types of rivers and mills—floating, overshot, etc.—is found in *Les Cartulaires des Templiers de Douzens*, ed. Pierre Gérard and Elisabeth Magnou, (Paris: Bibliothèque Nationale, 1965), intro. pp. xxix-xxxii. Interestingly enough, there is quite a clear reference here to the construction by the Templars of a fulling mill, "ad molendinum draparium edificare," document, no. A17 (1153), not known by Bantier, see n. 100 below.

⁹⁸ For example, *Cart. de Silvanès*, nos. 59 (1152), 176 (1149), 253 (1143), 267 (1151), 9 (1133).

The dilapidated condition of mills or weirs is implied, for example, when the existing mills at Camboulas, under mortgage, were conveyed to Bonnecombe, *Cart. de Bonnecombe*, no. 273 A (1193).

⁹⁹ Numbers of sites for mills acquired by Cistercian houses in southern France can be tabulated thus:

Abbey:	Sites with Existing Mills:	Sites for New Mills:	Totals:
Berdoues	9	3	12
Bonnecombe	12	0	12
Bonnefont	3	4	7
Bonnevaux	6	1	7
Boulbonne	8	1	9
Candeil	5	2	7
Fontfroide	4	1	5
Franquevaux	7	—	7
Gimont	11	3	14
Grandselve	5	11	16
Léoncel	1	—	1
Silvanès	2	2	4
Valmagne	7	1	8
Other Abbeys	16	0	16
Totals	96	29	125

plans for the construction of "two mills: one for fulling and the other for grinding under one roof" at Promillac; this is one of the earliest documented fulling mills in southern France.[100] There are other indications of similarly elaborate milling complexes built by the order, including floating mills on the Garonne River built by Grandselve, and the large complex with at least six mills at Paulhan on the Hérault River owned by Valmagne.[101] Despite the enormous interest which Cistercians took in water mills in southern France, however, crediting the order with the introduction of water-powered mills in this region is a mistake; what the order did in many cases, was to improve or maintain existing mills.[102] Indeed, about two-thirds of the rights granted to the order for mills and milling complexes were at sites which already had one or more millwheels.[103]

Closely aligned to capital investment in mills and mill construction was irrigation or drainage of meadows near those mills, and the creation and stocking of ponds to raise fish. Improvements to meadows in southern France generally meant either constructing ditches to drain marshes and convert them into meadows, or construction of facilities to flood meadows periodically for increased hay production. The order's emphasis on animal husbandry meant that such meadows were particularly valuable to the Cistercians, since they supplied the additional hay which made it possible to keep and feed livestock over the winter. Cistercian irrigation of meadows in conjunction with the order's transhumance, discussed in the next chapter,

---

[100] *Cart. de Silvanès*, nos. 210 (1159) and 145 (1164). No. 210 reads, in part, "et damus vobis paxeriam ad prata facienda in Promilac, et aliam paxeriam ad molendina facienda unum ad molendum et alterum, si vobis placuerit, ad parandum, et duo piscatoria in capitibus pauxeriae, sed in molendinis nemo alius debet molere vel parare preter vos et si commutaveritis vel dederitis alicui predicta molendina vel si aliquis extraneus ibi molerit vel paraverit preter vos, nos habebimus molturam et paraturam." In terms of industrial production, the mill for fulling planned by Silvanès may represent a contribution by the new religious orders; see Anne-Marie Bautier, "Les Plus Anciennes Mentions de moulins hydrauliques industriels et de moulins à vent," *Bulletin philologique et historique (jusqu'à 1610) du Comité des Travaux Historiques et Scientifiques* (1960): 1: 567–625.

[101] For example, floating mills in the Garonne belonging to Grandselve: Paris, B.N. Coll. Doat, vol. 76, Grandselve, no. 57 (1198), rights to rebuild a mill at the chateau of Pin granted to the owner of that chateau, who must nonetheless do so without damage to Grandselve's mill at Cabanac: Toulouse, A.D. Haute-Garonne, 108H 56 Grandselve, no. 277 (1205), mills at the meeting of the Garonne and Nadesse rivers, ibid., no. 251 (1241), and Montpellier, A. D. Herault, film "Cart. de Valmagne," vol. 1, fols. 135v–37r (1174), in which Guy Guerriatus, son of the William of Montpellier who died as a monk at Grandselve, gave the mills at Paullan to Valmagne.

[102] It is not as Jean Gimpel, *The Medieval Machine* (New York: Penguin, 1976), 3–5, implies when he describes Cistercian mills as factories with each grange of 742 abbeys in the order having identical plans in the twelfth century. Not only is he exaggerating the number of twelfth-century Cistercian houses (there were about 500), but the implication that these abbeys were responsible for the introduction of such mills cannot be upheld.

[103] At a number of sites the monks had multiple mills at a single emplacement or dam; for example, Paris, B. N., Coll. Doat, Grandselve, vol. 76, fol. 17r (1188), Paris, A. N. L1009 bis, Grandselve, no. 135 (1186), Paris, B. N., Latin MS 11008, Grandselve, fol. 12v (1181). That the number of milling emplacements often exceeded the number of granges that an abbey had, suggests that many of these mills were operated for their seigneurial proceeds.

enabled the land to support more livestock per acre, which, in turn improved land fertility, for there was more animal manure available to fertilize the fields. Drainage and watering of meadows are mentioned in a number of documents for abbeys in the region. Fontfroide's inventories record at least four donations or purchases of rights to construct ditches, presumably for irrigation.[104] A Silvanès document specifically mentions rights to "make a meadow."[105] Bonnevaux charters include a document which outlines how water will be divided among Bonnevaux and six lay proprietors *ad prata irriganda*.[106] Similarly, meadows conveyed to Gimont were presumably for haymaking (by irrigation?).[107] Also, fish ponds were invariably constructed behind mill dams to supply protein for the monastic diet of fast days and during Lent, but there is no indication that ponds were also intended to produce fish for market.[108]

In addition to mill upkeep, construction, and related water works, capital improvements to Cistercian granges definitely included construction of strong barns at the granges, which are well-documented from the end of the thirteenth century, and certainly deserve considerable additional study by archaeologists and historians of domestic architecture.[109] The order's willingness to invest in large capital improvements such as mills and barns, and its general concern with labor-saving technique and equipment, are probably important reflections of other Cistercian management practices. If Cistercians expended funds for such high-priced items as mills and barns, they probably also invested in smaller agricultural improvements: better plows, hand-tools, and other implements, as well as improved seed, and draft and breeding animals. All these last items, however, leave little trace in the surviving records, which concern primarily only land acquisitions.[110] It is important to note that capital investment in mills, like Cistercian acquisition of tithes, is not an index of the order's tendency to seek seigneurial rights; except in acquisition of such mills and tithes, the monks rarely sought or obtained the *bannal* or seigneurial rights so often associated with rural lordship or manorialism. For example, concessions of ovens to the order

---

[104] Carcassonne, A. D. Aude, MS H211, "Inv. de Fonfroide," fol. 87 (1245).

[105] *Cart. de Silvanès*, no. 227 (1164): "ad prata irriganda."

[106] *Cart. de Bonnevaux*, nos. 212, 215, and 288 (all undated).

[107] *Cart. de Gimont*, II, no. 209 (1222) and VI, no. 62 (1164).

[108] Permission to build fishponds is given specifically in *Rec. de Bonnefont*, no. 130 (1178) and in Paris, B. N., Coll. Doat, vol. 83, Boulbonne, no. 65 (1212). Cistercians like other Benedictines did not eat the meat of quadrupeds, thus fish was important for their diet and fishponds were normally constructed; see Lekai, *Cistercians*, 318–19. Rights were also acquired for coastal fishing, for example, by Valmagne: Montpellier, A. D. Hérault, film "Cart. de Valmagne," vol. I, fols. 137v (1148), 138r (1168), 138r (1171).

[109] On Cistercian barns, see Berman, "Fortified Granges," passim. Little evidence from earlier than the fourteenth century is available on such structures.

[110] Much of the evidence for the introduction of the tools described by Lynn White, Jr., *Medieval Technology and Social Change* (New York: Oxford University Press, 1962), 39–78, comes from manuscript illuminations, architectural sculpture, and painting, sources not usually available for Cistercian houses, although there are several early manuscripts from Cîteaux showing Cistercians using simple tools: see Dimier, *L'art cistercien*, 325.

seem to have been limited to those for the abbeys' own use; similarly, concessions of judicial rights over villagers were explicitly made to the monks only in one or two cases.[111]

If it represents anything beyond a search for self-sufficiency in the rural world, the order's acquisition of mills in rural areas, and eventually in urban areas as well, must be considered a response to or reflection of its general interest in labor-saving in its economic activities. Eventually this would lead to the development of water power for industrial uses in the conversion of raw materials such as wheat, wool, bark and forest products, or leather into semi-processed goods—a development which is found primarily for daughter-houses of Clairvaux like Grandselve, and primarily after the mid-thirteenth century.[112] Initially, however, the building and improvement of mills and barns, the purchase of tithes, and other practices verging on seigneurialism must all be viewed as reflecting the needs of the order, in its introduction of grange agriculture into settled areas, or of capital investment, rather than as lapses from the order's ideals which forbade ownership of such tithes or profiting from seigneurialism.

A final area of possible improvement in agricultural yields on monastic estates was new agricultural techniques or practices: new crop rotations, marling, spreading of additional animal manure on the fields or grazing animals on them, introduction of new varieties of seed, and so on. Even more than for capital improvements, however, there is little available evidence of such innovation. It is also doubtful that sophisticated techniques of crop rotation, which begin to be documented for Cistercians in northern France in the twelfth and thirteenth centuries, were also introduced by Cistercians in southern France.[113] The "old-fashioned" two-year rotation

---

[111] On ovens, see *Rec. de Bonnefont*, no. 341 (1246), and Paris, B. N., Coll. Doat, vol. 84, Boulbonne, fol. 54 ff. (1231). Justice in villages was acquired late, when at all; see *Cart. de Bonnecombe*, no. 51 (1244), for the conveyance of such rights to Bonnecombe: "Notum . . . quod nos frater P, abbas Bonecumbe, consensu et voluntate conventus ejusdem domus, confitemur et recognoscimus coram vobis domino B. de Galliaco, bajulo Amiliavi et quarumdam partium Ruthenensis diocesis pro dicto domino comite Tholosano, nos velle teneri ab hac die et in antea libere majus dominium quod habemus in villa de Coms vel habere possemus vel debemus et sine aliquo servitio et sine aliquibus exactionibus a dominio comite Tholosano personaliter quamdiu vixerit."

[112] Paris, B. N., Coll. Doat, vol. 114, nos. 55 (1208) and 112 (1192), referring to *moulins drapiers et tanniers* and *moulins bladiers* acquired by Candeil, is one example. See also, Mireille Mousnier, "L'abbaye cistercienne de Grandselve du XIIe au début du XIVe siècle," *Cîteaux: comm. cist.* 34 (1983), 220–44.

[113] The introduction of the three-field system is described in White, *Medieval Technology*, 69–76; however, the problem is much more complicated than White thought: see Duby, *Rural Economy*, 94–96; Douglass C. North and Robert Paul Thomas, *The Rise of the Western World: a New Economic History* (Cambridge: Cambridge University Press, 1973), 41–43; and Slicher van Bath, *Agrarian History*, 20. On the three-field system used by Cistercians in northern France, see Higounet, *Vaulerent*, 41–46. The Ile-de-France, however, where the monks of the Cistercian house of Chaalis, as Higounet noted, were using the three-field rotation system, was a region in which that system had long been used: see Duby, *Rural Economy*, 22–25. North and Thomas stress the need for introducing such rotation only where a high density of population exists, denying the validity of claims that it constituted an absolute improvement in agricultural technique no matter where introduced.

was generally considered the more suitable rotational pattern for agriculture in Mediterranean regions than the new three-field rotation used in the medieval north of France;[114] a charter from Gascony indicates that this older two-year rotation was in effect when Cistercians acquired land there.[115] If additional harvests were wrested from the fields by the order in southern France, it is likely to have been through the introduction of an occasional extra crop of such marketable produce as dyestuffs rather than by a change in the cereal rotation.

Whatever increase in cereal production on Cistercian granges resulted from improvements in technique, it is clear that the introduction of a larger number of animals per cultivated acre increased soil fertility. Increased numbers of livestock, as discussed in chapter 5, would have allowed Cistercian farming to break out of the vicious cycle associated with traditional medieval agriculture, in which poor yields resulting from insufficient animal fertilizer forced peasants to turn waste into arable land in order to feed human populations, thus decreasing the number of animals grazing on the waste. Consistent reduction of animals per farm or per acre of cultivated land meant that there were fewer draft-animals and therefore plowing and harrowing were less efficiently done, that animal protein in the peasant diet was low, that peasants lived almost completely on cereals and suffered the lethargy of malnutrition, and that animal fertilizer continued to be insufficient and tended to be used only on gardens, vineyards, and other hand-cultivated areas.[116] An increase in the number of the order's animals through transhumance and the production of hay improved this situation and broke the cycle.

Cistercian cereal production probably took up more acres of cultivable land than did the peasant agriculture which had preceded it, since Cistercians did not cultivate gardens to the extent that peasants did, but the order did not limit its activities altogether to cereal production or stock-raising. Most abbeys also owned vines, and it is clear that Cistercians in southern France, like their Benedictine predecessors, attempted to be self-sufficient in wine production. Gifts of vines and vineyards were made to almost every house in the order, and the largest abbeys, Grandselve, Bonnecombe, and Valmagne, in particular, each had properties devoted exclusively to

---

[114] Slicher van Bath explains that the two crop rotation was used in the south because in Mediterranean areas winter wheat would have been the principal cereal, since sparse spring rainfall made it difficult to grow spring cereals: *Agrarian History*, 59.

[115] *Cart. de Berdoues*, no. 409 (1189): "Sciendum est quod Gillelmus de Lavedan, prior de Sancto Oriento cum consilio et voluntate sui conventus . . . miserunt in pignus Arnaldo abbati Berdonarum et conventui ejusdem loci, . . . omnia jura que habebant vel habere debebant in ecclesia de Sancto Orientio in decimis et primiciis terrarum culturarum et incultarum, que fratres Berdonarum tenent et possident sive pro dono sive pro pignore sive pro emptione in grangia de Sparciaco ante Solsonum et retro infra Ercium et seram de Begaut, . . . pro CL solidis morlanorum . . . Hoc pignus debet stare per tres bladationes que faciunt sex annos. . . . M.C.LXXX.IX." Not surprisingly such a mortgage contract is more explicit about such practice, since there is a question of who gets the fruits of the land, and when.

[116] Slicher van Bath, *Agrarian History*, 7–18.

wine production.[117] Eventually Grandselve must have produced considerable amounts of wine to be shipped down the Garonne for sale in Bordeaux.[118] In general, however, there is little indication that most southern French Cistercian houses actively sought to expand their practice of viticulture and wine production beyond internal needs. Vineyards, whether they consisted of strips of vines adjoining cereal-producing granges or were located in centers separate from other cereal-producing farms (like Bonnecombe's holdings at Bougaunes), remained limited. Although holdings of vines and vineyards were sometimes called "granges," may have been worked by lay brothers, and had tithe exemptions, they were not as efficient as the cereal-producing estates, because they were much more labor-intensive.[119]

It was for the cereal-producing estates that the primary advantages of Cistercian grange agriculture were felt in southern France. These large, compact granges created by Cistercian monks in southern France, worked by *conversi* (who were often assisted by the monks in the early years), were probably the major source of profits to the order once they were operating. Even without production figures, it is clear that Cistercian granges were more profitable than the holdings which had preceded them and that such granges considerably improved *net* yields, allowing the production of surpluses for sale in the growing towns of the region. The advantages of Cistercian grange agriculture discussed here, as they apply to the Cistercian economy in southern France, were basically the result of a dramatic reorganization of the cereal-producing lands acquired by the monks. This amelioration of southern French agriculture, however, is not attributable to "new" lands, or "bumper crops," or "virgin" soils, or "pioneering" activity, but stemmed from the order's managerial efficiency. If in addition to making Cistercians rich and supplying nearby towns, these management practices provided a model (and sometimes the cash) for neighboring peasants and lords to improve their own agriculture, that would have been an added benefit to the region's economic growth.

---

[117] Grandselve leased out its vine strips, in Toulouse, A. D. Haute-Garonne, 108 H 58, Grandselve, 45 (1242), 310 (1169), 301 (1169), 43 (1243), etc. Grandselve had vineyards associated with its properties in Toulouse: Toulouse A. D. Haute-Garonne, 108 H, liasse 5 (1213–14, etc.); Valmagne had vineyards near the Hérault, Montpellier, A. D. Hérault, film, "Cart. de Valmagne," vol. 2, fols. 114v–42r.

Bonnecombe's cellar, house, and other equipment at Bougaunes; Rodez, Soc. des lettres, MS Bonnecombe, "Cart. d'Iz et Bougaunes," nos. 47 (1181), 91 (1191), 102 (1191).

[118] Grandselve had rights to transport wine on the Garonne as far as Bordeaux: Paris, B. N., Latin MS 11010, Grandselve, passim, and Mousnier, "L'abbaye," pp. 220, 231 ff.

[119] Duby, *Rural Society*, 139, discusses the need for intensive labor in viticulture. Often gifts of vineyards were given for commemorative purposes. For example, Villelongue received three *arpents* of vines from Regina and her six sons for the soul of her husband and their father, buried by the Cistercians: Paris, B. N., Coll. Doat, Villelongue, vol. 70, no. 24 (1157).

V. PASTORALISM AND TRANSHUMANCE

It has already been argued that Cistercian pastoralism in many cases must have provided the capital for the earliest Cistercian land acquisitions, and that an increase in the number of animals per acre under Cistercian cultivation improved land fertility and probably allowed the order's cereal cultivation to break out of the vicious cycle of low yields typical of the medieval period. These results were primarily because Cistercians in southern France acquired considerable amounts of pastureland beyond that associated with their agricultural lands. Such additional lands obviously added further to their growing wealth. For some abbeys, animal husbandry probably remained the most important source of cash, and the relative effectiveness of different houses in instituting large-scale pastoralism was certainly one cause for growing differences in wealth and power which arose among the order's abbeys in southern France. Pastoralism also reflected the order's interest in economic pursuits which were not labor intensive, since even from the earliest years, labor recruitment was difficult for many abbeys, as discussed above.

Although there was sometimes cooperation between mother- and daughter-abbeys or those within a particular filiation, more often pastoralism was a source of conflict. Pasture rights belonging to smaller houses in the order were often annexed by their more powerful neighbors; even more in jeopardy were pasture rights belonging to the affiliated houses of nuns. Controversy also arose between Cistercians and other religious houses, particularly with the Templars, Hospitallers, and Praemonstratensian canons. Surviving Cistercian archives supply plentiful documentation of the order's early and continuing interest in acquiring pasture rights. Records of arbitrated settlement of these quarrels, both within the Cistercian order and with its neighbors, provide considerable evidence about the location of Cistercian pastures and the order's practice of transhumance. The settlements of the disputes were often recorded in agreements called *convenientiae* which give detailed descriptions of pasture rights claimed or ceded.[1] As a result of such settlements, certain abbeys' pasture rights were limited to strictly-defined areas, but in those areas abbeys were sometimes also allowed to monopolize resource use. In some agreements maximum numbers of animals to be allowed into areas of pasture are mentioned, and such documents also provide information about the passage of Cistercian

---

[1] On these contracts, see Paul Ourliac, "La 'Convenientia,' " *Etudes d'histoire du droit privé offertes à P. Petot* (Paris: Montchrestien, 1959), 413–22.

flocks and herds, as well as recording the boundaries of areas in which different abbeys were granted rights to pasture their animals.

The *convenientiae*, as evidence of Cistercian pastoral practice, are supplemented by large numbers of actual conveyances of pasture rights to the order by individual donors. Such donations were often made in particularly vague terms, as grants of pasture in the territories held by particular donors, but their location can sometimes be determined by the toponymic surnames which the donors used to identify themselves as principals in the charters.[2] These charters document pasture acquisitions by the order in areas near Cistercian granges and rights to pasture in addition to those normally associated with the arable land which the order acquired. Charters also conveyed rights for the passage of animals through pasturelands belonging to other Cistercian houses, or for the "ascent to and descent from the *montagnes*" of animals belonging to the monks.[3] Exemptions from passage tolls granted by various lords also help to pinpoint the routes used by Cistercians in their transhumance.

Access to extensive pasture allowed Cistercians to acquire and feed very large flocks and herds. Unfortunately, it is rarely possible to give actual numbers for those animals or to "quantify" pastoralism, except by citing the extent of pasture donations, or by arguing that the controversy which Cistercian pasture use began to arouse in the early thirteenth century is itself witness to the extensiveness of Cistercian flocks and herds. A few acts give numbers for specific geographical areas; the Silvanès cartulary also sometimes indicates what animals were involved.[4] In one charter for that abbey, rights to browse in the forest were allowed for a number of the abbey's pigs.[5] Another document for Silvanès reveals that at Calmels, a pasture area on the eastern part of the Causse de Larzac, the monks could keep four pair of oxen, thirty horses and a dozen cows with their young, and a flock of sheep.[6] Although that act does not say how many

---

[2] "Pasture in all my land or territory" references are found in *Cart. de Bonnecombe*, nos. 180 (1206), 181 (1207), 188 (1214), 200 (1220), 201 (1223), 202 (1223); no. 209 (1231), etc. The last says: "M.CC.XXX.I. Ego Bego Ysarnz de Mirabello, . . . dono et concedo Domino Deo et Beate Marie Bonecumbe et Willermo abbati et vobis fratribus ejusdem loci presentibus et futuris omnes herbas meas atque herbatges et omnes aquas et omnia nemora mea ad usus vestros et omnium peccorum vestrorum, jumentorum, et animalium, ut libere et quiete habeatis et perpetuo jure possideatis."

[3] Paris, B. N., Coll. Doat, vol. 76, Grandselve, no. 48 (1178); Paris, B. N., Coll. Doat, vol. 138, Bonnecombe, no. 9X (1242) (for the quote) and in the vernacular charters phrases like "passar, apojar et adavallar": see *Cart. de Bonnecombe*, no. 217 (1245).

[4] Animals are often mentioned in the contracts; for example, conveyances of land to the order in return for animals or animal products include at least six horses (*roncini, caballi*, etc.) used as payment by Silvanès (*Cart. de Silvanès*, nos. 199 [1158], 47 [1139], 326 [1155], 244 [1167], 245 [1169], 276 [1154]), and over fifty sheepskins or fleeces used in payment for land purchases by that same abbey. The numbers mentioned are discussed below, but they rarely concern the animals held by the largest abbeys in the region.

[5] *Cart. de Silvanes*, no. 297 (1152).

[6] Ibid., no. 381 (1147): "M.C.XL.VII., ego Raimundus Ricardi de Foderia et ego Jordanus, frater ejus, . . . vendimus, donamus atque laudamus cum hac presenti carta monasterio Salvaniensi, in honore Dei genitricis semperque virginis Marie constructo, et tibi Guiraldo, abbati,

sheep were there, other evidence suggests that a normal Cistercian flock must have been about one thousand or twelve hundred animals. One document for Silvanès limited the sheep to be held in the parish of Serruz to a thousand.[7] Similarly, the Cistercians of Valmagne were allowed to keep a thousand sheep in the territory of Omellaz.[8] In Provence, Léoncel's flock in one area was limited to twelve hundred animals.[9] Although there are no numbers given for the size of Fontfroide's flocks in the county of Foix, the Roussillon, and in Spain, its total holdings in livestock must have been similar in numbers to those of its daughter-house Poblet; the latter in 1316 had 40 horses, 111 cattle, 2,215 sheep, 1,500 goats, and 172 pigs.[10]

In developing animal husbandry Cistercians profited from the economic trends of the twelfth and thirteenth centuries which favored producers of animal products over those of cereals. The demographic expansion of the High Middle Ages, accompanied as it was by the growth of towns and urban industries and by migration into those towns from rural areas, was characterized by a rise in demand for the necessities of life—food, clothing, and shelter—but also by an improved living standard. The average diet, previously based almost entirely on cereals, began to be more varied—with meat and other animal proteins replacing some of the cereal consumed. Markets for leather, woolens, and parchment also expanded. Thus, as the structure of demand changed, cereal prices stagnated while demand for more costly food items—meat, cheese, and butter, and for clothing (made

---

. . . pascua, herbarum in toto hoc quod habemus et habere debemus in territorio de Calmelsz. . . . Animalia vero que in predicto territorio alere et tenere debetis sunt hec: grex ovium unus, boum quatuor paria, eque XXX, vacce XII, ita quod pulli et vituli earum ibidem morantes cum matribus infra unum annum in prescripto numero non computabuntur. Sciendum vero est ut quocienscumque de numero predictorum animalium, vaccarum videlicet sive boum quicquam minus fuerit ex vaccis sive bubus complebitur. Item notum sit ut si forte quandoque in prefato territorio hominum habitationem fieri contigerit, et ipsa habitatio ultra tres focos excreverit, non sicut ante sic postea animalia vestra passim et ubique per territorium pascentur, sed erunt pascua eorum usque ad terminos quos talibus exprimimus vocabulis." See also ibid., no. 177 (1153).

[7] Ibid., no. 70 (1153): "M.C.LIII . . . Ego Bertrandus de Ponte et ego Guillelmus et ego Aimericus et ego Berengerius et ego Arnaudus, nos omnes fratres . . . donamus et laudamus et confirmamus et titulo donationis cum hac carta tradimus monasterio beate Marie, quod vocatur Salvanesc, et tibi domno Guiraldo, abbati. . . . et donamus et laudamus pascuas ad alenda omnia animalia que modo tenetis vel in antea tenebitis in Grauson, in tota parrochia de Fragos, et in sursum usque ad Cellers et per la Calm et in Domeiran et in Lobeira de strata in intus usque ad Carrairolam; et ad omnia animalia, que modo tenetis vel in antea tenebitis in Gallac, sicut descendit serra usque ad Sanctum Martinum et sicut venit per la Calm usque ad Campum Revel; et ad omnia animalia que modo tenetis vel in antea tenebitis in Cantalops vel in Campolongo et in monasterio, et ad unum tropellum de ovibus, in tota parrochia de Serruz, quod tropellum non debet excedere numerum mille. . . . donamus et concedimus vobis ligna sicca in tota terra nostra ad omnes usos vestros necessaria."

[8] Montpellier, A. D., Hérault, film "Cart. de Valmagne," vol. 2, fols. 151r (1187) and 172v–73r (1210).

[9] *Cart. de Léoncel*, no. 26 (1163–69): "Notum . . . quod dominus Arbertus de Turre donavit pascua per omnem terram suam, et filii ejus Arbertus et Berlio, excepto mandamento de Turre, ad quadraginta trigenarios ovium." But see *Charles d'Aiguebelle* no. 115 (1295).

[10] Narbonne, B. Mun. "Inv. de Fonfroide," fol. 37 (1316), cited in Grèzes-Rueff, "Fontfroide," 271.

from wool and leather) increased. This favored producers like the Cistercians, who were willing and able to invest in stock-raising.[11] Cistercian development of pastoralism, however, was probably not a conscious response to these market conditions. The order developed pastoralism primarily because it suited its needs. It was through pastoralism that Cistercians most successfully detached themselves from human society. The sylvan isolation of Cistercian pastoralism recalled that of the eremitical groups who had often preceded the new order at its sites. For those who sought life "far from human habitation," pastoralism must have been particularly appealing, and that life-style was, after all, what had attracted many new members to the Cistercian order.[12]

Probably, one reason why the order's abbeys acquired the reputation for pioneering activities in the wildernesses of medieval Europe was that its pasture rights were often acquired in forested areas.[13] This was certainly true in southern France. Early conveyances of pasture rights to the monks there often included numbers of concessions allowing pigs and other animals to feed in woodland.[14] Such concessions in forest were often included in more general grants of forest rights or usage: to collect fallen wood, to gather berries and honey, to cut branches for animal bedding, and to hunt game.[15] Such "usage" might include explicit rights to cut dead trees for fuel or a specific number of live trees for building projects—for example, a certain number of oaks annually to keep mills in repair, to build scaffolding or make roof timbers, or to fashion wooden tools.[16] Concessions of "forest usage" generally imply and sometimes explicitly state, however, that clearance and conversion to arable land were not to be conducted by the monks

---

[11] Duby, *Early Growth*, 141–43, and Robert H. Bautier, *The Economic Development of Medieval Europe*, trans. Heather Karolyi (London: Harcourt, Brace and Jovanovich, 1971), 117–18.

[12] Pastoralism had been practiced by eremitical groups like the Carthusians; the tendency for stock-raising to bring wealth, however, was probably the reason for its rejection by the Grandmontines: see Lackner, *Eleventh-Century Background of Cîteaux* (Washington, D.C.: Consortium, 1972), 201, and Jean Becquet, "L'érémitisme clérical et laïque dans l'ouest de la France," *Eremitismo in Occidente nei secoli XI e XII*. Settimana internazionale di studio: Mendola, 1962, (Milan, 1965), 27–44. Unlike the Carthusians, Cistercians did act as shepherds to their own animals. Pastoral imagery is frequent in Cistercian writing: see for example Bernard of Clairvaux's *De Consideratione*, trans. John D. Anderson and Elizabeth T. Kennan (Kalamazoo, Mich.: Cistercian Studies, 1976), 2: 15, 4: 3, 4: 6.

[13] This has often been described as sylvo-pastoralism: see Duby, *La Ferté*: "Selon toute apparence, les religieux de la Ferté furent des pasteurs beaucoup plus que des défricheurs; entre 1113 et 1178 ils semblent bien avoir fondé essentiellement l'économie de leur domaine sur l'utilisation par l'élevage d'un patrimoine laissé à la forêt et à la solitude"—intro., 18–19.

[14] *Cart. de Silvanès*, nos. 297 (1152) and 145 (1164).

[15] For example, *Cart. de Gimont*, II, no. 102 (1207): in which Gimont received rights in a church: "cum omnibus terris cultis et incultis in ejusdem ecclesie territorio consistentibus, cum ingressibus et egressibus, aquis, pascuis, et nemoribus, et cum omnibus ad venatum pertinentibus."

[16] *Cart. de Berdoues*, nos. 97 (1167), 567 (1215), and 697 (1256)—all concessions of rights to cut building materials; Paris, A. N. L1009 bis, Grandselve, no. 29 (1165); and Paris, B. N., Coll. Doat, vol. 76, Grandselve, fol. 139r (1198).

or their laborers.[17] Examples of conveyances of forest usage almost inevitably included pasture rights, whether mentioned explicitly or not. For instance, pasture rights were certainly part of the concessions to Berdoues of usage in the forest of Mated or that of Ozalt, and pasture was probably included along with rights to collect firewood in the territory of St. Martin of Tels; its usage in forested areas was explicit in the concession of pasture, which included wood for building *cabanes* (shepherds' huts) in that area.[18] Similarly, among charters for Valmagne are conveyances to that abbey of rights in the forest of Burau with pasture usage in the parish of St. John of Fraissenet: "in waters and woodlands, in waterways or in meadows, in empty or occupied areas, in lands which are cultivated or uncultivated."[19] Grants of rights in the woods of Chevremorte and Serreneuve for firewood and building materials and other "usage" were similarly made to the monks of the abbey of Villelongue.[20] Grandselve received general forest usage rights in the *nemus* of Gothmerlenc, in that of Nonas, and that of Rabastens.[21] In the *nemus* of Calcantus it received wood for making charcoal as well as rights to annual *glandage* (for pigs to feed on nuts and acorns).[22] Grandselve also received forest usage in the *nemus* of Candeil, where it eventually founded its daughter-house of that name.[23] The monks of Candeil were later also granted rights there as well as in the woods of Peirola and of Lavaur.[24] Such concessions almost invariably included rights to take animals into wooded areas to feed, at least at certain times of year. These examples are but a few of a much larger number of such conveyances of forest usage near their granges to Cistercians in southern France, on which the order's reputation for sylvo-pastoralism was based.

Cistercian abbeys also acquired more general rights to pasture for their animals in the wastelands and even the fallow lands belonging to their neighbors. Such concessions described by donors in the charters as "pasture

---

[17] *Cart. de Berdoues.* no. 411 (1186): "Tamen in nemore de Gavere non debent predicti fratres facere novale sine consensu eorum . . ." or a document for Valcroissant stating: "Nullatenus debent extirpare vel rumpare vel alio modo destruere nemore, que tempore dicte concessionis et statuti census erant propria Dyensi ecclesie et ecclesie Valliscrescentis, vel aliquam partem dictaturum nemorum, faciendo mortalia vel esarta, vel ex alia quacumque causa." Jules Chevalier, *L'abbaye de Notre-Dame de Valcroissant de l'ordre de Cîteaux au diocèse de Die* (Grenoble: Allier, 1897), 157.

[18] *Cart. de Berdoues*, no. 288 (1188), "totum nemus quod vocatur Mated et omnem espleitam exinde habentem de Ozalt," and ibid., no. 328 (1185), "Espleitam arborum ad opus cabanarum et ignium," (*cabanes* were, of course, shepherd's huts).

[19] Montpellier, A. D. Hérault, film, "Cart. de Valmagne," vol. 1, fol. 84v (1170): "Totum honorem habemus in totam quartam partem in forestibus de Burano et de Sauze et de Bescabes . . . in aquis vel in nemoribus sive in aquarum cursus vel decursus sive in pratis vel in heremis vel condrictis in terris cultis vel incultis . . . in parrochia de Sancti Johanni de Fraissento."

[20] Paris, B. N., Coll. Doat, vol. 70, Villelongue, nos. 19 (1152) and 21 (1153).

[21] Paris, B. N., Coll. Doat, vol. 76, Grandselve, nos. 6 (1188), 62 (1198), and 75 (1204).

[22] Paris, B. N., Latin MS 11011, Grandselve, nos. 95 (1165) and 71 (1166) and n. 27 below.

[23] Paris, B. N., Coll. Doat, vol. 76, Grandselve, no. 8 (1150).

[24] Paris, B. N., Coll. Doat, vol. 114, Candeil, no. 69 (1177), no. 79 (1188), and no. 90 (1189).

in all my lands" were frequently added to the end of contracts of land purchase; they have the initial appearance of a casual afterthought. Only their reappearance again and again in the contracts indicates the diligence with which pasture was sought by the monks. Indeed, there are frequently recorded conveyances of pasture by "donors" who neither gave nor sold any other rights to the monks.[25] As a result, Cistercians acquired pasture rights not only in areas where the order cultivated land, but also in distant territory. This assiduous collection of such pasture rights allowed the monks to expand their herds beyond the numbers which could be supported on their own granges. In some areas these rights were to land at elevations where agriculture was impossible because of the short growing season or because terrain was too rough or dry, but gifts of pasture also were in areas cultivated by neighboring lords and peasants. The result of these concessions was that Cistercians' access to pasture was far in excess of what would normally have been associated with their land, and the total number of animals that the order could raise increased considerably, as well as increasing relative to acreages devoted to cereal cultivation.[26] This allowed monastic agriculture to escape the cycle of infertility associated with too few animals per acre, which so often plagued medieval agriculture.

The order would not have continued to seek out new rights to pasture, or to the toll exemptions which allowed abbeys to move animals to market without payment, if pastoralism had not been practical as well as profitable. With regard to practicality, most decisive at the outset was the fact that the Cistercian intensification of pastoralism in southern France had lower start-up costs and was less labor-intensive than cereal cultivation or viticulture—a situation particularly important in the early years when Cistercian houses had limited economic resources and personnel. Moreover, resources needed for the flocks and the access to pasture, water, and rights for passage of animals to pasturelands, were often given without cost as outright gifts, to the monks.[27] This meant that abbeys had only to provide

---

[25] Toulouse, A. D. Haute-Garonne. 108 H 56, Grandselve inventory, passim; entire sections of archives and cartularies are devoted to pasture rights: see, for example, *Cart. de Berdoues*, nos. 267–373, and *Cart. de Bonnecombe*, nos. 264–398, which are based on separate bundles of donations of pasture rights.

[26] Increases in effective pasture were also made by acquiring and exploiting hay-meadows— for instance *Cart. de Silvanès*, no. 37 (1133)—or by irrigating them so they would produce more hay; see pp. 89–90.

[27] Most conveyances were made without cost to the monks: see Paris, B. N. Latin MS 11008, Grandselve, nos. 58 (1164), 77 (1165), 101 (1171), 94 (1168), 147 (1174), 166 (1180), 936 (1181), 940 (1182), etc., which are land conveyances with concessions of pasture added to them. Berdoues acquired pasture rights from most of the leaders of the nearby community soon after its foundation: see *Cart. de Berdoues*, nos. 353 (1243), 371 (1145), 364 (1248), and 268 (1157). The most expensive acquisition of pasture rights by the order may have been that made by Berdoues from Arnold of St. Roman in 1245 who granted the monks pasture rights as surety for a loan of 400 sol., *Cart. de Berdoues* no. 371 (1245): "Arnaldus Sancti Romani miles et dominus d'Ornezan . . . gratis et spontanea voluntate . . . donavit et concessit . . . Deo et beate Marie Berdonarum et Hugoni abbati . . . pasturas, herbagia, introitum et exitum liberum sine lesione peiarum et fossarum et omnem espleitam necessarium per omnes terras

themselves with stock.[28] Also adding to the attractiveness of pastoralism was the fact that the papally-granted tithe exemption included stock-raising.[29] The exemption from tithes which would otherwise have taken every tenth new-born animal was a stimulus to the growth of the order's flocks and herds, allowing more animals to be kept and bred, sold for cash, or traded for new land. Without a tithe collector taking his choice of the best of their animals, the monks could more easily select the best animals for breeding and more quickly improve the quality of their flocks and herds.[30] That Cistercian animal husbandry was tithe-exempt increased still further the net profits of stock-raising and breeding for the order. Animals could be raised at relatively low cost and could be marketed by the order free of taxes and tolls. Indeed, even if the monks adhered to the injunction of the Benedictine rule, "to sell at lower prices than their neighbors," their profits could easily have been higher than those of their competitors.[31]

In conveying pasture to the Cistercians during the twelfth century, the laity was often quite heedless in making grants without limits or safeguards to protect themselves.[32] There are several reasons why this may have been so. First, concessions of pasture appear to have been made by local men because the abbeys were poor and their flocks and herds so insignificant that they would in no way deprive a donor's animals using the same pasturelands. Second, it is possible that some donors of pasture sought to profit from the folding of sheep or grazing of other animals belonging to the monks on their land, or that they granted such pasture in order that their own animals have access to the order's breeding animals.[33] Finally, pasture rights in southern France may not yet have become a commodity to be bought and sold like land, but continued to be resources considered limitless—to be conceded to the monks by lay owners at will, without any

---

suas, videlicet glandem et faginam, et omnes fructus cujuscumque sint generis, aquas, et nemora, et omnia que sunt pastoribus et animalibus cujuscumque generis sint necessaria, . . . per omnes terras suas . . . ad utilitatem dictorum fratrum Berdonarum exceptis terris bladatis, ortis, et vineis cultis, . . . M.CC.XL.V." In most cases, however, pasture was conveyed *gratis*.

[28] Bonnecombe was given flocks and herds in a bequest in the will of Raymond VI of Toulouse (see *Cart. de Bonnecombe*, no. 216 [1242]), but this was unusual.

[29] The exemption from tithes on Cistercian stock-raising is usually expressed along with that on cereal cultivation under direct management of the monks, as in the papal privilege of 1178 for Bonnecombe from Alexander III—"Sane laborum vestrorum, quos propriis manibus aut sumptibus colitis sive de nutrimentis vestrorum animalium, nullus omnino a vobis decimas vel primitias exigere presumat"—*Cart. de Bonnecombe*, no. 64 (1178).

[30] On Cistercian selective breeding, see Robert Fossier, "L'essor économique de Clairvaux," *Bernard de Clairvaux*, préface de Thomas Merton. Commission de l'histoire de Cîteaux no. 3 (Paris: Alsatia, 1953), 111.

[31] *The Rule of St. Benedict*, trans. Anthony C. Meisel and M. L. del Mastro (Garden City, N.Y.: Doubleday, 1975), 93, chap. 57.

[32] Many donors gave gifts similar to one to Gimont of, "pascua, et herbagges et liberum introitum et exitum per omnes terras suas": *Cart. de Gimont*, VI, no. 125 (1190).

[33] Folding sheep is discussed in detail by Hill, *English Cistercian Monasteries*, pp. 75–77. For southern France, the only explicit document is a very late agreement between the monks of Aiguebelle and villagers who required animals to be kept every night on their cultivated lands; *Chartes d'Aiguebelle*, no. 115 (1295).

diminution of value and no anticipated restriction on the donor's own continued use. This rapidly changed. The casual attitude regarding pasture concessions to the monks continued only until the late twelfth century when Cistercian exploitation had intensified to such a degree that laymen were forced to become more wary of alienating their pasture rights unconditionally to the monks.

Although Cistercians, like other users, had always been expected to pay for any damage done by their animals to standing crops, vines, or meadows,[34] thirteenth-century donors of pasture began to impose stringent limits on the extent of the Cistercians' pasture use. Sons and grandsons of those who had earlier conveyed unlimited pasture rights to the order began attempting to limit its access to pasturelands.[35] Disputes soon arose between more aggressive abbeys and their lay neighbors over the allocation of certain pasture resources, but by the thirteenth century, Cistercian access to such resources was well established. Although the heirs of the original donors complained about excessive use of pasturelands by the monks, they succeeded only in limiting, not in excluding the order's animals altogether. For example, in one thirteenth-century contract recording the settlement of a dispute, Bernard of Lévézou managed to confine Bonnecombe's pasture rights in the region of the Rouergue called the Lévézou located southeast of the abbey, to an area west of the Muse River.[36] In a similar document of the 1240s concerning territory north of Bonnecombe, Bertrand of Balaguier is seen attempting to preserve pasture near Rignac for his own use and that of his men at Mirabel, by limiting the number of Cistercian-owned animals there. Although he was not able to deny Cistercian access altogether, the contract guarantees it to him and his men.[37] Similar contracts

---

[34] Some of such gifts also included the proviso, "exceptis terris bladatis, horis et vineis cultis," *Cart. de Gimont* VI, no. 125 (1190), but limitations on usage are almost nonexistent in twelfth-century conveyances of pasture to the monks.

[35] See *Cart. de Bonnecombe*, no. 214 (1240), "Item dixit et voluit sepedictus arbiter quod prefatum capitulum presens et futurum habeat absque contradictione dicti Archambaldi et ejusdem posteritatis et possideat in perpetuum pacifice et quiete universas herbas et pascua animalibus ejusdem capituli necessaria et usus nemorum et aquarum in toto et ex toto territorio, mandamento, sive honore dicti castelli Petrebrune et castelli de Toeils sepedicti, salvis et exceptis necessariis usibus herbarum ad propria animalia Archambaldi et suorum hominum in dicta terra sua manentium sustentanda"; nos. 211 (1232); and no. 213 (1237): "et retento nobis nostro proprio usu et hominum nostrorum in herbis et pascuis supradictis."

[36] Ibid., no. 212 (1233): "M.CC.XXX.III. II. ydus julii, cum controversia verteretur inter Poncium de Mandalazac monachum et fratrem D. Paliz et R. de Salis et U., fratres domus Bonecumbe, nomine ipsius domus ex una parte, et nobilem virum Bernardum de Levezo ex altera, et peterent dicti fratres nomine monasterii sui erbas et erbatges et pascua et aquas et nemora ad usus hominum et animalium suorum, pecorum, jumentorum et armentorum suorum per totam terram dicti Bernardi de Levezo et specialiter per totum territorium de Levezo usque ad rivum Muse, quia dicebant quod pater dicti B. et mater sua et ipsemet Bernardus donaverant et concesserant Deo et Beate Marie Bonecumbe."

[37] Ibid., no. 217 (1245): "Sentence arbitrale entre l'abbé de Bonecombe et Bertrand de Balaguier, au sujet de droits de pacage" in the vernacular: "Volem e comandam e dizem jutgan que la maios de Bonacomba aja e tenha d'aissi enan totas las herbas els pasturals, senes contradizemen d'en Bertran de Balaguier, que so entr'Also e Biaur els boscs e las aigas que auram ops al bestial de la dicha maio et als homes quel bestial capdellarau."

established in the thirteenth century limited Cistercian usage elsewhere, but generally they could not altogether override the effects of early grants.

The extensive grants of valuable pasture rights made to Cistercian houses throughout southern France in the twelfth century were crucial in the order's early expansion, but in developing that pastoralism the monks did not confine themselves to collecting access to grass and water for their animals. Other rights equally valuable to Cistercian pastoralism were the exemptions from passage tolls, which allowed them to move their animals as well as other goods over long distances to markets without payment, and grants of the salt essential to the diet of the animals, as well as for preserving animal products like butter, meat, and cheese. Salt was usually given to an abbey as a set amount to be collected from a salt-works each year; often rights to transport that salt without toll were also acquired. Such grants came from the highest authorities. Grandselve received from the kings of England and dukes of Aquitaine the grant of a boatload of salt annually from salt-works at Bordeaux. Additional rents in salt were given to Grandselve by the lords of Lesparre and the viscounts of Narbonne.[38] Silvanès, in addition to a gift of ten mule-loads of salt annually from the archbishop of Narbonne, received both an annuity in salt and the source of that salt: a mound of sea-salt and presumably the associated salt-works, after the death of the donor, Berengar Rotbald of Capestang.[39] Bonnefont was granted a *sestier* of salt weekly by one donor, and Valmagne, Franquevaux, and probably Boulbonne as well, were granted rights to operate their own salt pans along the Mediterranean.[40] Such gifts of salt-works or of annual rents in salt contributed considerably to the viability of Cistercian pastoralism. The amounts of salt needed by Grandselve, for instance, also suggest the scale of the order's stock-raising.

The amount of capital tied up in Cistercian animal husbandry in southern France is probably best demonstrated by reference to the agreements by

---

[38] Paris, B. N., Latin MS 11010, Grandselve, nos. 18 (n.d.), 50 (n.d.), 59 (1199), and 62 (1218); and Latin MS 9994, Grandselve, no. 746 (1143).

[39] *Cart. de Silvanès*, no. 400 (1159): "M.C.L.VIIII. . . . Ego Berengerius, Narbonensium archiepiscopus et Apostolice Sedis Legatus, . . . dono et laudo . . . beate Marie de Salvanesc et tibi Guiraldo, . . . ut de vestris propriis causis domus vestre nullam leddam nec ullum usaticum nec ullam omnino consuetudinem in Narbona donetis. . . . Similiter ego predictus Berengerius, . . . dono vobis predictis fratribus Salvaniensis monasterii in unoquoque anno apud Sanctum Georgium X saumadas de sale ad onus saumariorum vestorum" and no. 403 (1167): "M.C.LX.VII., ego Berengerius Rotbaldi de Capite Stagno, . . . dono et laudo monasterio beate Marie de Salvanesc et tibi Poncio, abbati, et fratribus ejusdem loci, ut apud Caput Stagnum in meo podio vestrum sal ponatis absque ullo vestro usatico; et dono vobis in unoquoque anno unam saumadam de sale in vita mea tantum. Post mortem vero meam relinquo vobis unum montem salis."

[40] *Rec. de Bonnefont*, no. 91 (1169), Montpellier, A. D. Hérault, film, "Cart. de Valmagne," vol. I, fol. 142v (1177): "unum scampnum salinarium in terminii Sancti Martini de Caucz quod est de XXXVIII salinas." *Layettes du Trésor des Chartes*, ed. Alexandre Teulet et al. (Paris: Plon, 1863–1909), 3: no. 3578, p. 2, mentions Boulbonne's claims to pass with its salt free of toll; Nîmes, A. D. Gard, H33 "Inv. de Franquevaux," fol. 161r (1192), and Paris, B. N. Coll. Doat, Belleperche, vol. 91, no. 29 (1236), and Dupont, "Exploitation du sel," 14.

which Bonnecombe bought a grange from Bonneval in the 1220s. The charters say that for that grange Bonnecombe paid 10,000 *solidi*, not in cash, but in live animals; to Mazan were promised four bulls valued at 5,000 *solidi*, to be transferred at the fall livestock fair in Rodez, and to Bonneval were paid eighty-four head of cattle also worth 5,000 *solidi*. Later, another 2,000 *solidi* were paid to Bonneval when the documents were transferred.[41] In addition to showing the large sums of Cistercian capital tied up in animals, this settlement suggests the profits to be gained from pastoralism.

In addition to expanding their pastoralism into a wide area beyond the parishes in which their abbeys and granges were located, Cistercians in southern France were successful in instituting the regular practice of the semi-nomadic, seasonal movement of animals from mountain to plain and back again, which is called transhumance. This practice, which had the effect of maximizing use of rich grass growth in summer pastures in the mountains and in winter along the coasts, and which allowed the monks to maintain larger numbers of stock, contributed considerably to the order's economic success.[42] The practice of transhumance was not a Cistercian innovation in southern France, and the order's shepherds were neither the first nor the only ones to take advantage of the region's potential for seasonal migration of flocks and herds. Transhumance had been used by individual peasants in areas of Provence for centuries and Benedictine houses had probably adopted the practice long before the twelfth century.[43] Moreover, during the twelfth century, other new religious orders competed with the order in practicing transhumance. Nonetheless, although the Cistercians were not the only new monks practicing such pastoralism, those monks developed it into a much more complex and efficient movement of animals

---

[41] *Cart. de Bonneval*, nos. 140–42bis (1225) and no. 151 (1232).

[42] European transhumance is described by Fernand Braudel, *The Mediterranean and the Mediterranean World in the Age of Philip II*, trans. Sian Reynolds (New York: Harper, 1972), 1: 88 ff.

[43] For Provence in particular, see Thérèse Sclafert, *Cultures en Haute-Provence: Déboisements et pâturages au moyen age* (Paris: S.E.V.P.E.N., 1959), 3: "Dans la montagne, l'homme ne peut vivre que s'il a des troupeaux, et les troupeaux ne sont vraiment rentables que si l'on peut, l'hiver, les nourrir hors du terroir, dans les plaines ensoleillées. D'autre part, les troupeaux du bas-pays ne peuvent se maintenir que s'ils trouvent dans la montagne les pâturages de l'été. Or la Provence possédait, dans les hautes vallées du nord et de l'est, les montagnes couvertes de fraîches pâtures qui pouvaient accueillir le bétail pendant les mois d'été; au sud, dans les chaudes et basses vallées, les troupeaux venus du nord trouvaient une vaste zone d'herbage qui leur assurait la nourriture d'hiver." Ibid., 11–24, describes the transhumant activities of the monks of St. Victor of Marseille, as well as of independent peasants. See also, Henri Cavailles, *La Vie pastorale et agricole dans les Pyrenées des Gaves, de l'Ardour et des Nestes* (Paris: Université de Paris, 1931), passim; Max Sorre, "Etudes sur la transhumance dans la région montpellieraine," *Société languedocienne de géographie*, Bulletin 35 (1912): 1–40; Jacques Bousquet, "Les Origines de la transhumance en Rouergue," *L'Aubrac: Etudes ethnologique, linguistique, agronomique et économique d'un établissement humain* (Paris: C.N.R.S., 1971), 2: 217–55; and Pierre Coste, "La vie pastorale en Provence au milieu du xive siècle," *Etudes rurales* 46 (1972): 61–75.

than did their lay neighbors.[44] The practice of such pastoralism must have presented problems of internal discipline for Cistercians, since it meant either hiring reliable outsiders as shepherds, or having monks and *conversi* involved in long treks out of the immediate control of their superior, buying and selling animals at fairs, and being part of legal proceedings over pasture rights or damages done by their animals.[45] Such disruptions, however, were probably outweighed by the profitability of transhumance. In southern France, on the other hand, Cistercian transhumance involved shorter distances and was more modest in scale than that practiced elsewhere in the Mediterranean; there were no uprooting of whole villages, disruption of family life, and governmental interference as found in the transhumance practiced in areas of Spain and southern Italy.[46]

The numbers of conveyances, exemptions from tolls, the frequent controversies, and so on, suggest the large scale of Cistercian transhumant pastoralism in southern France. For, whatever the advantages in terms of health of sheep or increased animals per acre, the most important reason that transhumance was practiced by Cistercian abbeys in that region was that they could not have supported the large numbers of animals they owned without it. Acreages near most abbeys would have supported a much smaller number of animals year-round. By sending animals into the mountains in summer, producing hay to feed them through the winter, or moving them into the mild winter climate of the marshes on the Mediterranean coast, Cistercians expanded their pastoralism beyond the limits imposed by the months of meager grass production on their own lands. Thus, increased numbers of animals were possible with transhumance because pasture in the northern Rouergue, for instance, could feed many more animals in summer than in winter, whereas pasture along the Languedoc coast could support those excess animals in winter, but had no grass for them in the hot, dry summer.

While it is difficult to provide numbers, it is possible to describe the areas in which Cistercian houses in southern France held extensive pasture rights and indicate the major orientations of their transhumance. The order's transhumant flocks and herds were moved by a variety of routes. Each abbey had its own group of pasture rights for summer and winter feeding, each had its own set of exemptions from passage tolls allowing its animals to be moved from winter pasturelands to the summer highlands—the *alps*, *montages*, or *estives*, and back. The movement of sheep and other animals

---

[44] As discussed below, competition with other religious groups was keen in the Rouergue; on Gascony, see *Cart. de Berdoues*, nos. 108 (1204), 91 (1258), 89 (1252), and 90 (1253).

[45] That some of these things became problems is obvious from complaints in the General Chapter meetings, see *Statuta*, ed. Canivez, "prima collectio," 1134, no. 51, "De nundinis." The Cistercians had more money, more political power, more time and longevity than the laity, and could carefully put together the complex sets of rights necessary to practice transhumance, whereas smaller castle-holders and peasants often could not.

[46] Compare the patterns described by Braudel, *Mediterranean*, 1: 91–4, and by Peter Partner in Italy, *The Papal States under Martin V* (London: British School of Rome, 1958), 18 ff.

was often quite gradual, following relatively short routes radiating upwards from the valleys of the Garonne and Rhône, or from the Languedoc coastal plain into a variety of elevated pasturelands. Often the animals followed special sheep-roads, walled expanses wider than any road and generally not coinciding with regular routes, which were called *drayas* or *drailles*.[47] In general, houses in Gascony and in the vicinity of Toulouse and Narbonne had pasture rights in the highlands of the French Pyrenees. Houses in Provence used pasture in the Alps or alpine regions along the coast, but also moved animals across the Rhône River to take them up into the Cévennes and Massif Central. The abbeys of the Rouergue, the Albigeois and those on the Languedoc plain concentrated their pastoralism on the mountain valleys and plateaux of the Cévennes and Massif. A more detailed look at such areas of pasture rights, as related to abbey and grange sites reveals how varied were Cistercian patterns of transhumance. See Maps 4 and 5.

As already mentioned, the Alps were an important area of pastoralism for Cistercian abbeys in southern France, and particularly for those in the marquisate of Provence. Pasture rights in that region stretched from the Viennois and Valentinois eastward, and Cistercians apparently moved large numbers of sheep and other animals from winter pastures along the Rhône into the Alpine meadows. Pasture rights for Léoncel, for example, stretched from the confluence of the Isère and Rhône Rivers southward to the Drôme, and then east of the Rhône for at least forty kilometers.[48] Beyond the pasture areas to which Léoncel's animals had access were those belonging to the Cistercians of Valcroissant.[49] Slightly northward, Bonnevaux had even more extensive pasture rights described in numerous donations of pasture made by men of all levels of society in that region. Surviving documents for that abbey include an agreement between Bonnevaux's monks and the religious order of la Grande Chartreuse, which shows that Bonnevaux pastured its animals in a region stretching from the Rhône up into the Alps north of Grenoble.[50] Disputes over pasture in the region east of the Rhône occurred

---

[47] See Bousquet, *Transhumance*, 233–37.
[48] *Cart. de Léoncel*, nos. 23 (1173), 24 (1174), 27 (1178), 34 (1185), 35 (1185), 42 (1191), 50 (1193), etc.
[49] Few documents survive for this abbey, but see Chevalier, *Valcroissant*, passim.
[50] *Cart. de Bonnevaux*, nos. 429 (ca. 1130) and 430 (1185). The latter is an arbitrated agreement with the house of la Grande Chartreuse. A confirmation of pasture rights by the count of Vienne, ibid., no. 431 (1222) shows that those Cistercians had rights stretching to very near the Carthusian house and other religious communities: "Ego Andreas Delphinus, Albionis et Vienne comes, approbo et confirmo omnes donaciones quas domina ducissa mater mea omnesque predecessores mei domui Bonevallis usque ad hanc diem donaverunt. In primis concedo et confirmo donum quod Guigo comes et Matildis uxor ejusdem eidem domui dederunt, scilicet quandam contaminam que est juxta aquam Velchiam, in parrochis Sancti Saturnini, et percursum et pascua per totum comitatum Albionensem. Item confirmo et concedo predicte domui alpem illem que dicitur Chalmencus, ad estivandas oves suas, ita quod nullus alius in eisdem pascuis animalia sua ad pascendum possit inducere; que scilicet alpis terminata est, in presentia Matildis regine et domini Johannis abbatis Bonevallis primi, ab oriente a quadam rupe que vocatur Berad, sicut aqua pendet usque ad rivum Rosset, a meridie descendit usque

as frequently as elsewhere; one document indicates that lay neighbors of the Cistercians were accusing the order of encroaching on their pasturelands.[51] Such conflict may even have denied certain Cistercian houses access to the Alps, for the monks of Aiguebelle seem to have had no rights in that area, but instead took their flocks across the Rhône into the Massif Central, where their pastoralism came into conflict with that of the Cistercians at Mazan in the Vivarais and with the abbey of Bonneval in the northern Rouergue.[52]

From the limited surviving documents it appears that Bonnefont and l'Escaledieu, which were sited in the Pyrenees midway between *estives* and plain, practiced a transhumance which extended from their granges in Gascony up into the highest Pyrenees and across the divides into Spain.[53] L'Escaledieu held pasture in the valleys near its original abbey site high in the Pyrenees at Campan, from which its monks had moved in the twelfth century to a more clement location.[54] Bonnefont, with its granges in the Comminges, had pasture nearby in the Ariège, but also summer meadows farther south in the high valleys of the Neste, the Pique, and the Rioumajou Rivers.[55] Both abbeys had daughter-houses beyond the Pyrenees in the northern parts of *reconquista* Spain.[56] Their abbots crossed that divide annually on visitation, and their shepherds could easily have gone into Spain with flocks and herds.[57] Later references confirm that this was also the case for animals belonging to Fontfroide, located near Narbonne in Languedoc, but with pasture rights reaching south into Catalonia.[58]

Transhumance is more difficult to document in the charters of the two most important abbeys of central Gascony—Gimont and Berdoues. Their documents, although recording considerable acquisition of pasture rights within the region of these abbeys' granges, provide no direct evidence of

---

ad stratam que exit a Cartusia et vadit ad castrum Cornilionem, quam stratam sequitur ab occidente usque ad ponticulos et ascendit usque ad rupem que vocatur Estreit, et pervenit usque ad terram fratrum Calisiensium et inde venit usque ad monticulum qui vocatur Coche."

[51] *Cart. de Léoncel*, no. 79 (1216).

[52] *Chartes de'Aiguebelle*, nos. 44 (1215), 47 (1217), 49 (1219), 56 (1234), and 58 (1235).

[53] Locations are described by Gratien LeBlanc, "La réparitition géographique des abbayes cisterciennes du sud-ouest," *France méridionale et pays ibériques; mélanges géographiques offerts en hommage à M. Daniel Faucher*, (Toulouse: Privat, 1949), 2: 584–608.

[54] Granges for l'Escaledieu, for which no charters survive, are indicated by locations of *bastides* later founded on them: see Higounet, "Cisterciens et bastides," *Paysages*, 265–74; LeBlanc, "Répartition," 585–89.

[55] *Rec. de Bonnefont*, ed. Samaran and Higounet, intro., 31.

[56] See Higounet, "Une carte des relations monastiques transpyrénéennes au Moyen Age," *Revue de Comminges* 64 (1951): 128–38, on the ease of crossing the Pyrenees during the Middle Ages. Daughter houses of l'Escaledieu and Bonnefont south of the Pyrennes were Fitero, Monsalut, Oliva, Sacramenia, Veruela, and Bugeo for l'Escaledieu, and la Baix and Sante Fé for Bonnefont; see Janauschek, *Originum*, 47–48, 65, 67, 91, 119, 167–68, 226, 227.

[57] A point underlined also by Emmanuel LeRoy Ladurie, *Montaillou: The Promised Land of Error*, trans. Barbara Bray (New York, 1979).

[58] Grèzes-Rueff, *Fontfroide*, 271.

transhumance. Although this may be explained as a *lacuna* in the surviving documents for Gimont, it is otherwise for Berdoues.[59] Numerous charters concerning pasture are recorded; indeed, they form an entire section of the Berdoues cartulary, but no specific acts concern areas recognizable as mountain pastures or *estives*, nor are there passage rights for flocks and herds recorded. It is possible that Berdoues shared pasture in the Pyrenees with other houses in the filiation of Morimond, such as l'Escaledieu, and that Gimont did likewise, rather than acquiring their own rights to summer pasture, but such cooperation is not documented.[60] Given Cistercian practice elsewhere in the south, it is less likely that Berdoues and Gimont did not practice transhumance at all.

Although several abbeys located in the middle Garonne valley in the region surrounding Toulouse had rights to summer pasture in the high elevations of the Ariège and eastern Pyrenees, it was Grandselve's rights there which were most extensive. Much of the evidence for Grandselve's extensive network of pasture rights is found in the *convenientiae* or arbitrated settlements of disputes over pasture and passage, which established boundaries between the abbey's exclusive areas of pasture in the eastern Pyrenees and those of rival houses. Concessions of pasture in the mountain valleys of the Ariège or exemption from tolls on moving its animals to those summer pasture areas were granted to Grandselve by a variety of high political authorities: among others, the count of Toulouse and his vassals or rivals in Foix, Comminges, Armagnac, and Béarn.[61] These concessions by the greatest territorial lords of the region were supplemented by those of lesser men who actually controlled access to particular valleys or tolls on routes leading to those places.[62]

Grandselve's pasture rights, however, began in the vicinity of the abbey itself, near Toulouse; even there, arbitrated settlements provide information on boundaries established between pasture areas reserved to Grandselve or its neighbors. For the area of Castelsarrazin, between the Tarn and the Garonne Rivers, boundaries were established in 1174 by the abbot of Clairvaux between areas of exclusive pasture for the Cistercian abbey of Bel-

---

[59] Gimont's cartulary ends rather abruptly and the Doat Commission did not gain access to its later charters in the seventeenth century. It is possible that other bundles or volumes of acts in that abbey's archives concerned pasture rights exclusively: see H. Omont, "La Collection Doat à la Bibliothèque Nationale," *Bibliothèque de l'Ecole de Chartes* 77 (1916): 292.

[60] A close relationship between l'Escaledieu and Berdoues dated to the participation of l'Escaledieu's first abbot in the foundation of Berdoues: see *Cart. de Berdoues*, no. 125 (n.d.), and discussion in Berman, "Growth."

[61] Paris, B. N., Grandselve, Latin MS 9994, fols. 221v (1143) and 262v (1186); Coll. Doat, Grandselve, vol. 76, nos. 52 (1195), 83 (1208), and 86 (n.d.) and vol. 77, no. 98 (1217); Toulouse, A. D. Haute-Garonne, Grandselve, 108 H 56, nos. 133 (n.d.), 120 (1225), and 108 H 46, no. 88 (1180).

[62] For example, Sicard Alaman, who although untitled was an official of the count of Toulouse, controlled tolls in the region of Albi: see Charles Higounet, "Les Alamans, seigneurs bastidors et péagers du XIIIe siècle," *Annales du Midi* 65 (1953): 227–54.

leperche and Grandselve's areas of pasture rights.[63] Similarly a contract between Grandselve and the Cistercians of Gimont used the Serrampione River as the boundary between pasture areas for the two houses.[64] Donations, as well as such arbitrated settlements, also show that Grandselve had extensive rights on both sides of the Garonne and Hers Rivers, stretching eastward up into the Albigeois.[65] Southwest and west of the abbey it had pasture near its grange or stock farm at Larra.[66]

Grandselve's mountain pasture areas, however, were probably most im-

---

[63] Paris, B. N. Coll. Doat, vol. 76, Grandselve, no. 46 (1174) and vol. 77, no. 107 (1227), in which the abbot of Berdoues resolved additional differences between those two houses over pasture: "De omnibus controversiis et querimoniis, et malefactis, que erant inter domun Grandissilve et domum Gemundi pro pascuis que sunt inter Petronum de Siraco at abbatiam Gemundi, miserunt se in manu Vitalis abbatis Berdonarum . . . praedictam controversiam pacificaverent in hunc modum constituerunt siquidem ut pascua sint communia domui Grandissilve et domui Gemundi que sunt a supradicto Petrono de Siraco usque ad Ardisas, et de Ardisas usque ad Insulum et ab Insula sicut ascendit Sava usque ad molinarium de Marestanho et iterum de dicto Petrono versus Cumbas Dausimpodio et inde usque in Gimona et sicut ipsa Gimona ascendit usque in ponte Castri Sancti Joannis Veteris, et sicut ascendit rivus qui est iuxta pontem usque ad metas positas in via veteri que exit a castro Sancti Joannis usque ad Caminum Sancti Jacobi, quod venit a domo monialium Sancti Joannis et vadit ad Pontem Romevium et a ponte Romevio versus Hospitalem Amboni, et ab hospitale Amboni usque ad Hospitalem de Bestiol, et ab hospitale de Bestiol in directum versus castrum Montisferrani, et inde versus Gaillard Viela et inde per serram usque ad supradictum molendinum de Marestanho, quod est in Sava. Et est sciendum quod animalia Grandissilve possunt transire metas positas iuxta rivum usque in serra que est a Sancto Joanne usque in via publica Sancti Jacobi. Similiter constituimus ut animalia Grandissilve domus possint adaquari et bibere infra terminos praedictos ubicumque voluerint, ad mensuram superioris vadi quod est iuxta pontem Sancti Joannis veteris, mensura autem illa talis est, ascendit supra montem usque ad fontem et a superiori parte usque ad Rivum de Scotacaval [?], et ab inferiori parte tantum quantum ascendit usque ad rivum supradictum et caetera adaquatoria habeant eandem mensuram, sane quia de honore de Folhaco exierat dissentio et controversia institutum est, ne unquam de caetera domus Gemundi vel domus Grandissilve possit ibi aedificare, nec alicui dare vel vendere, vel impignorare, aut aliquomodo ad aedificandum alienare, sed utraque domus ibi habeat proprietates suas absque damno alterii domus, et utraque pars illorum non possit ibi mittere bestiarium infirmum vel extraneum in praedictis pascuis, sed utraque pars defendat praedicta pascua ab hominibus omnibus bona fide et predictas pascuas teneant et explectent salvis vigiliis cabanarum utriusque domus cum omni bestiaro suo cuiuscumque modi, vel generis sit, habeant inde glande et folia e ligna et omnia necessaria. . . . De aliis autem pascuis que habent ultra Ras et in Lomania, nullus alterii faciat controversiam, vel impedimentum vel calumniam . . . et quod domus Gemundi videbatur esse aggravata pro predictis pascuis et terminis, domus Grandissilve pro bono pacis dedit domui Gemundi centum libras morlanorum bonorum, de quibus fratres ipsius domus tenent se pro bene contentis . . . Anno ab incarnatione Domini millesimo ducentisimo vicesimo septimo."

[64] *Cart. de Gimont*, III, no. 50 (1158): "Sciendum est quod abbas Grandissilve absolvit abbati Gemundi omnem querelam grangie de Francavila et grangie del Forc, et abbas Gemundi absolvit abbati Grandissilve omnem querimoniam grangie de Tarrida, hoc modo. . . . Pascua ita divisa sunt: terminus pascuum est la comba que descendit d'Auzimpoi ad fontem de Reginar, qui descendit ad Montel qui est juxta Sarrampion et est terminus adhuc ille rivus qui exit de Sarrampion denant lo Montel et vadit versus Ardizan versus orientem . . . M.C.L.VIII."

[65] Paris, B. N., Latin MS 9994, Grandselve, fol. 282r (1190): "Ego Pontius Amaneus . . . dono et concedo in perpetuum deo et Beate Marie Grandissilve et Willelmo abbati et fratribus eisdem loci presentibus et futuris pascuam animalibus eorum in omni honore meo quem habeo et habere debeo inter Tarnum et Garonnam ut libere pascant per totum exceptis blatis et vineis et prata defessa . . . M.C.XC," and Latin MS 11008, Grandselve, fols. 102v–104v (1186).

[66] See Paris, B. N. Latin MS 11008, Grandselve fols. 116 rff. for Larra.

portant. Each year in early summer the monks of Grandselve moved their animals southward into the pastures of the Ariège and Pyrenees, where the abbey held rights to *estives* known as the *montagnes de Ravat*, between Tarasçon and St. Gaudens. To get the animals into this area of the Pyrenees, Grandselve took them through lands south of Toulouse where pasture was controlled by its daughter-house of Calers, the Cistercians of Boulbonne, and the regular canons of Combelongue. *Convenientiae* between Grandselve, Boulbonne, and Combelongue established that each house could move animals through the lower Ariège on their way up into regions of exclusive summer pasture in late spring, and come back again through those areas in the fall.

Convenientiae also established rights in the mountain valleys themselves. One such agreement made with the canons of Combelongue, dated 1178, allowed that in coming and going from Ravat, the shepherds of Grandselve could move animals through areas where the canons of Combelongue normally had exclusive pasture rights; there the monks could rest, feed, and give water to their animals for three days in the spring and for three days when they descended again in the fall.[67] It allowed both Grandselve and Combelongue exclusive pasture within certain areas of the *estives*, establishing boundaries between those two areas. In addition, the two houses held an area of common pasture there for their animals, although in the common area Grandselve was forbidden to build huts for its shepherds. A third region was delineated in which Grandselve's shepherds could take animals for water, but not allow them to stay and to feed. This last requirement suggests two things: that water was limited in the Pyrenees with access to it of high priority, and that flocks and herds like those of Grandselve were often continually on the move, because of the need for water as well as fresh grasses.[68]

By 1191 it had also become necessary for Grandselve to deal with the monks of Boulbonne regarding pasture rights in that same region of Ravat.[69] The contents of an agreement between Grandselve and Boulbonne in that year suggest that each side had trespassed on what the other believed its exclusive rights. Grandselve was required to move its animals through Boulbonne's area of exclusive pasture rights before mid-June in ascending to the mountain valleys or forfeit its access for the year. Also, through this arbitrated settlement, Grandselve acquired an increase in its areas of ex-

---

[67] Paris, B. N., Coll. Doat, vol. 76, Grandselve, fol. 143 ff., no. 48 (1178).

[68] Rights to *abreuvage* were often specifically mentioned. See *Cart. de Bonnecombe*, no. 197 (1218): "M.CC.XVIII. . . . Ego Frotardus de Turre et Gillelma, uxor mea, et Gillelmus et Frotardus, filii mei, . . . donamus, dimittimus; desamparamus Domino Deo et Beate Marie Bonecumbe et Amblardo abbati et fratribus ejusdem domus presentibus et futuris, omnes herbas nostras atque herbatges et omnia nemora nostra et omnes aquas nostras, ubicumque ea in presenti habemus aut in futuro quoquo modo habituri sumus, quatinus ad usus vestros et omnium peccorum vestrorum, jumentorum et animalium libere et quiete habeatis, teneatis, et jure perhenni possideatis."

[69] Paris, B. N., Coll. Doat, vol. 83, Boulbonne, no. 99 (1191).

clusive pasture rights in the *estives* of Ravat, at the cost of giving up properties held in the town of Tarasçon-sur-Ariège.

By recourse to arbitration, despite its considerably smaller size, Grandselve's daughter-house of Calers also appears to have been able to retain pasture rights in the upper Ariège and eastern Pyrenees in competition with other larger abbeys. In a settlement in 1239 with Boulbonne, the monks of Calers gained an exclusive area of pasture west of the road to Toulouse—an area in which Boulbonne's flocks were not allowed without permission from Calers, except for one day in spring and one in fall, while *en route* to Ravat.[70] Calers in turn had permission to pass through areas of Boulbonne's exclusive pasture rights for two days each way in going to and from its *montagnes* at Savardun, an area of pasture located between the Ariège and Hers Rivers to the east of the *estives* of Ravat.

The abbey of Boulbonne held rights west of Tarasçon-sur-Ariège, as revealed in the *convenientia* of 1191 with Grandselve which has already been mentioned, as well as those described in the agreement with Calers. It also had pasture east of the Ariège River, mentioned in an agreement dated 1166 made with the canons of Combelongue.[71] In addition to establishing boundaries for areas of exclusive pasturelands in the region of Savardun, this agreement required that the canons not found any new religious house in that region, presumably because such new houses would compete with Boulbonne's pasture rights and land holdings there. The smaller Cistercian houses of Nizors, Bouillas, and Flaran, all daughters of Bonnefont located in the region south and west of Toulouse, also had pasture rights in the eastern Pyrenees, but how extensive their flocks and herds were is difficult to judge; there is reason to believe that most of their pasture rights were shared with the mother-abbey of Bonnefont.[72] The monks of Fontfroide also held access to pasture in the Ariège.[73]

Abbeys neither close enough to the Pyrenees to look southward for summer pasture nor so close to the Alps as to move their animals eastward turned to a third area: the Cévennes and the Massif Central. The Cévennes are a long range of mountains extending westward from the Rhône to divide Languedoc from the interior; their pasture areas or *montagnes* were used extensively by Cistercian abbeys of the coastal plain as well as those farther inland. For abbeys of the Languedoc, summer pastures in the Cévennes were to some extent supplemented by spring pasture in the *garrigues*, the dry, rocky outcroppings along the coast itself. Abbeys located in the interior of the Massif Central also practiced a transhumance which involved the Languedoc, moving animals from the interior of the Massif Central where they summered, to the coastal marshes for winter pasture.

Among Languedoçian monasteries, Fontfroide was perhaps most aggressive in its search for adequate pasturelands. In addition to pasture

---

[70] Paris, B. N., Coll. Doat, vol. 84, Boulbonne, no. 172 (1239).
[71] Paris, B. N., Coll. Doat, vol. 83, Boulbonne, no. 40 (1166).
[72] *Rec. de Bonnefont*, intro., 32.
[73] Grèzes-Rueff, "Fontfroide," 275–78.

rights in the Ariège granted by the counts of Foix, the monks of Fontfroide acquired considerable access to pasture in the Montagne Noire—the part of the Cévennes just north of Carcassonne.[74] There the abbey's flocks and shepherds must have competed with those of smaller Cistercian houses, like Villelongue, Ardorel, and Silvanès, all located in the Cévennes themselves, as well as with those of Candeil in the southern Albigeois.[75] As already mentioned, however, Fontfroide's most important pasture rights may well have been in Spain. In contrast, the Languedocian abbey of Valmagne's surviving documents suggest a more uni-directional movement of its animals northward into the *garrigues* in spring and then into the Cévennes, where its flocks and herds competed with those of Silvanès for pasture north of Béziers.[76] In fall and winter Valmagne then moved its animals down into the marshes of the Camargue where it shared pasture in the *silva* of Ulmet with Sauveréal, as well as with Franquevaux.[77] Intermediate stages between summer and winter pasturelands were Valmagne's granges near the abbey itself and the *garrigues* in the region just west of Montpellier. Franquevaux's pastoralism is less well documented, but this abbey in the Camargue did have rights to pasture in the Lubéron and *montagnes* of Sisteron east of the Rhône, as well as in the Camargue itself.[78]

The monastery of Silvanès, located in the Cévennes between the coastal plain and the Massif Central, had an optimal situation for short-distance transhumance, with access to pasture both on the Languedoc coastal plain and in various parts of the Cévennes and Massif. These latter pasture rights were located southwest of the abbey in the Cévennes near its grange of Margnès and east of the abbey near and on the Causse de Larzac.[79] Pasture rights on the Causse held by Silvanès and by the affiliated nuns of Nonenque, however, were gradually eroded by claims to exclusive pasture rights there made by houses of Templars on the Causse de Larzac at La Calvarie and La Couvertoirade.[80] Moreover, Silvanès's pastoralism was also blocked

---

[74] Ibid., maps on 270-71.
[75] On pasture rights for those smaller abbeys, see Paris, B. N., Coll. Doat, vol. 70, Villelongue, no. 9 (1150); Paris, B. N. Coll. Doat, vol. 114, Candeil, nos. 107 (1190) and 123bis (n.d.); and Albi, A. D. Tarn H 38 "Répertoire de Candeil," nos. 89 (1202), 98 (1231), and 141 (1251), and references to Silvanès in the next notes.
[76] Montpellier, A. D. Hérault, film, "Cart. de Valmagne," fols. 1r-54v, section "Garrigues," and rights related to pastoralism in vol. I cartulary section entitled "From Princes," ibid., vol. 1, fols. 149v-end; on Silvanès's nearby rights, see *Cart. de Silvanès*, no. 308 (1161).
[77] Montpellier, A. D. Hérault, film "Cart. de Valmagne," vol. 2, fols. 172v-73r (1210).
[78] Nîmes, A. D. Gard, H33 Franquevaux, fol. 177v, no. 2 (1185) for pasture east of the Rhône, and H66 no. 7 (1174), the gift of pasture in the *silva Godesca* from Bremund of Uzès, presumably in the vicinity of Uzès, west of the Rhône.
[79] Rights in the Cévennes were granted in *Cart. de Silvanès*, nos. 411 (1147), 386 (1139), and 308 (1161); nearer home Silvanès had rights in the parish of Camarès, in that of Serruz, etc.: ibid., nos. 89 (1163), 70 (1153), and 177 (1153). On the Causse de Larzac, Silvanès maintained only a grange at Sollies: ibid., no. 473 (1170).
[80] On the difficulties encountered by the Cistercian nuns in this regard, see Giselle Bourgeois, "Les granges et l'économie de l'abbaye de Nonenque au moyen âge," *Cîteaux: comm. cist.* 24 (1973): 139-60.

to the northwest by an expansion of pastoralism by the Cistercians of Bonnecombe.[81]

The history of Silvanès's loss of pasture rights to stronger competitors in the southern Rouergue is but a part of a larger record of pastoral expansionism and monopolies over access established by the wealthiest Cistercian houses of that province in league with the Templars, at the expense of smaller abbeys and convents. The events in that region may well reflect the pastoral aggrandizement of the more powerful Cistercian abbeys in other parts of southern France. This aggressive expansion on the part of the larger abbeys, meant that by the late twelfth century two Cistercian houses, Bonnecombe and Bonneval, along with the Templars of Ste. Eulalie, had divided pasture rights in the Rouergue for their own use, effectively pushing out all other religious houses as competitors. Their increasingly commercialized transhumance was rapidly becoming big business. These events are known partially because documents for the small Cistercian house of Locdieu in the western Rouergue, which lost in this process even more completely than did Silvanès, have survived among archives of the Templar commandery of Ste. Eulalie.[82] Included are charters concerning pasture rights and other land originally held by Locdieu which were alienated to the Templars. This transfer was accomplished not by Locdieu or with its assent, but by the Cistercians of Bonneval, during a period when Locdieu and its properties had been given into the control of Bonneval because of Locdieu's debts. In return for the desired rights to pasturelands on the Causse de Larzac, which had originally belonged to Locdieu, the Templars granted Bonneval an area of exclusive pasture north and west of that house, on the Aubrac plateau in the northern Rouergue.[83]

The circumstances of Locdieu's loss of pasture rights to Bonneval illustrate not only the power politics practiced by the larger Cistercian houses of the Rouergue and the problems with debt of the smallest houses, but some of the confusion about relationships within the Cistercian order which continued throughout the twelfth century. Locdieu had been an eremitical house, probably founded by Gerald of Salles; it had been incorporated by the Cistercians, as had its mother-abbey Dalon, into the filiation of Pontigny.[84] The incorporation of Locdieu in 1162 by the Cistercians is likely to have been sought after Locdieu was already in financial difficulties.

---

[81] As reflected by Silvanès's inability to acquire rights in the parish of Comps, where Bonnecombe had rights: *Cart. de Silvanès*, no. 218 (1161) and *Cart. de Bonnecombe*, nos. 269 D (1188), 271 G (1191), 273 E (1193), 278 B (1199), 283 A (1202), 285 B (1204), 288 A (1207), 293 A (1214), 298 A (1224), and 309 A (1247).

[82] Toulouse, A. D. Haute-Garonne, H series, Ordre de Malte, Ste. Eulalie; some of these are published in Brunel, ed., *Les Plus Anciennes Chartes*, nos. 32, 42, 60, 71, etc. After passing from the hands of Locdieu to those of Bonneval thence to the Templars of Ste. Eulalie, the documents concerning pasture and other rights originally belonging to Locdieu were transferred with Templar property to the Hospitallers when Philip IV disbanded the Templar order at the beginning of the fourteenth century.

[83] *Cart. de Bonneval*, nos. 41 (1181) and 76 (1189).

[84] Janauschek, *Originum*, 147–48.

Whether or not there was earlier debt and whether or not incorporation brought relief, Locdieu appears to have been seriously in debt by the next decade. If we can trust the *Gallia Christiana* account, in 1178 the abbot of Locdieu had ceded his abbey to that of Bonneval for the sum of 20,000 *solidi*, presumably to pay off his creditors.[85] According to Cistercian General Chapter minutes, it was only after a protest by Dalon to that chapter, in 1214, that Locdieu once more become independent of Bonneval.[86]

Although Locdieu regained its independence, Bonneval's handling of Locdieu's financial crisis did not result in putting the struggling abbey back on its feet, for Bonneval did not return all of Locdieu's properties to it. Instead Bonneval seems to have interpreted its task as the annexation of the smaller abbey (in the same way that other hermitages and priories had been annexed by the order); it mixed Locdieu's properties into its own patrimony, presumably treating the appropriated resources as payment for the debt of 20,000 *solidi*. Bonneval took Locdieu's land and pasture rights at la Roquette, the grange located in the center of that pasture area on the Aubrac where Bonneval had been granted exclusive rights by the Templars, and conceded exclusive pasture rights to the Templars on the Larzac (including pasture previously held by Locdieu) in return for concessions on the Aubrac.[87] This had unfortunate consequences not only for Locdieu, but for Silvanès and Nonenque, for the flocks and herds of all three smaller religious houses were gradually driven out of the Larzac.

Just as Bonneval had created exclusive pasture areas for itself in the northern Rouergue, the monks of Bonneval and Bonnecombe would divide between themselves access to pasture located between the Aveyron and Viaur Rivers in the central Rouergue. After years of bickering over rights there, agreements were reached which initially also included a third party, Bonneval's mother-abbey of Mazan. Although Mazan, located in the Vivarais east of the Rouergue, later confined its pastoralism to the eastern Cévennes and Rhône valley, at this point it still had access to pasture and perhaps one grange in the Rouergue, where its daughter-houses, Silvanès and Bonneval, were also located.[88] The records of the General Chapter suggest that disagreement over pasture rights between Mazan and Bonneval dated from at least 1194, but whether there was a temporary solution to

---

[85] *Gallia Christiana*, I: 263–64 regarding Locdieu's seventh abbot, "Arbertus monachus ipsius domus anno 1177 Kal. Decembris praegravatam. Hic a Dalon ensi domo eam separavit, domuique Pontiniacensi per annum conjunxit. Cum autem Pontiniacenses eam viderent pene destructam, arbitrarenturque sibi impossibile fore eam reparare, abbas egit ut domui Bonevallensi subjiceretur, consentientibus Ademare viro venerabili, primo abbate Bonevallensi, et Guillelmo ejus successore, a quo Aribertus accepti XX milia solidorum. Rexit fere per annos 4 dimisit abbatiam anno 4."

[86] *Statuta*, ed. Canivez, 1: 1212, no. 54, and 1214, no. 62.

[87] See Brunel, *Chartes*, nos. 71 (1152), 80 (1157), 107 (1165), and 111 (ca. 1166); *Cart. de Bonneval*, nos. 3 (1162), 41 (1181), 70 (1185), and 76 (1189), and intro., p. ix; and Bousquet, "Transhumance," 133–37, 233–37, 245–48.

[88] *Cart. de Bonnecombe*, no. 64 (1178), mentions "grangiam de Manso Dei cum pertinentiis suis"; does this belong to Bonnecombe or to Mazan?

their conflict at that time is not clear.[89] In 1216 the chapter appointed abbots to settle the disputes which had again arisen.[90] Their solution, as recorded in a *convenientia* dated 1217, was to divide pasture in the central Rouergue between the three abbeys of Bonneval, Bonnecombe, and Mazan,[91] but this did not resolve the problem. Mazan was again at odds with Locdieu in 1219 over pasture in the "western Rouergue."[92] Finally in 1225 the controversy was settled when Bonnecombe agreed to purchase Mazan's and Bonneval's pasture and other property in the central Rouergue including Bonneval's grange of la Serre. As mentioned above, the price for this sale, paid for with livestock, was 10,000 *solidi*, half paid to Bonneval and half to Mazan.[93] Along with its new grange, which it renamed Bonnefon, Bonnecombe gained exclusive pasture south of the Aveyron River. Haggling continued for another six years and was again noted in the General Chapter meeting of 1230.[94] In 1231, a final settlement was made when Bonnecombe agreed to concede an additional 2,000 *solidi* to Bonneval when the latter handed over the land titles for the 1225 sale.[95]

The results of controversy and negotiation were similar to those reached between Bonneval and Ste. Eulalie. By the beginning of the second quarter of the thirteenth century Bonnecombe controlled most access to pasture in the central Rouergue, where it had an area of exclusive pasture rights. These lay south of an east-west division on the Causse Comtal north of Rodez (where both Bonneval and Bonnecombe had granges) and included rights held without interference from Bonneval in the valleys of both the Aveyron and Viaur Rivers, with pasture lands stretching southward to the Tarn, from near Millau westward to Bonnecombe's grange of Bernac west of Albi. Pasture rights may even have extended to Toulouse, where Bonnecombe probably sold its livestock.[96] Similarly, the monks of Bonneval had no Cistercian rivals for pasture north of the divide on the Causse Comtal and Causse de Sévérac between the Aveyron and Lot Rivers north of Rodez; in addition, they had exclusive rights in pasture north of the Lot in the western half of the Aubrac Plateau which the abbey had acquired from the Templars. In the fall as they went southward from the Aubrac to Bonneval's granges in the central Rouergue, Bonneval's flocks could descend the *drayas* of the Boralde ravine near the abbey itself. From the central Rouergue, its animals were then moved across the Grands Causses to granges in the Cévennes near Anduze and into the Rhône valley to be

---

[89] See complaints in *Statuta*, ed. Canivez, 1: 1194, no. 35.
[90] Ibid., year 1216, no. 60.
[91] *Cart. de Bonneval*, no. 136 (1217).
[92] *Statuta*, ed. Canivez, 1: 1219, no. 50.
[93] *Cart. de Bonneval*, nos. 140–42bis (1225); Albi A. D. Tarn, H 38 "Répertoire de Candeil," no. 98 (1231).
[94] *Statuta*, ed. Canivez, 2: 1230, no. 30; 1231, no. 98.
[95] *Cart. de Bonneval*, no. 151 (1232). The documents survived the fire in the Bonneval archives because they were transferred to Bonnecombe's archives at the time of the sale; they are published among the Bonneval charters.
[96] *Cart. de Bonnecombe*, 345–430 and Bousquet, "Transhumance," 233–37.

pastured in the Camargue or sold in Beaucaire and St. Gilles.[97] Bonneval's access to pasture in the Cévennes and its ability to move flocks to market in the Rhône valley or to winter pastures in that area, were strengthened by its authority over a new daughter-abbey, that of les Chambons in the Vivarais, after 1214.[98]

This detailed description of Cistercian pastoralism in the Rouergue shows the ruthlessness with which pasture rights were sought, and also suggests the importance of filiation ties between mother- and daughter-houses, like Bonneval and Mazan, or Bonneval and les Chambons.[99] The parcelling out of Locdieu's pasture rights by Bonneval contributed directly to the development of pasture monopolies by Bonneval, Bonnecombe, and Ste. Eulalie in the central Rouergue. Exclusive control of access by these religious groups allowed them to practice what would become by the fourteenth century, if not earlier, an increasingly commercialized transhumance. Cistercian shepherds took large numbers of animals from winter pastures in the Quercy, and moved them up onto the Aubrac plateau, or by way of the central Rouergue towards the Lévézou, the Grands Causses to the east of that region, and the high mountains of the Vivarais.[100] The example of the Rouergue suggests the scale of Cistercian pastoralism throughout southern France, for these rivalries and conflicts over such rights were certainly not isolated.

Cistercian pastoralism by mid-thirteenth century had begun to be increasingly commercialized and disputes over the tithes on Cistercian animals began to arise at that time. Documents for the Alps show from an early date that the order began to commercialize pastoralism by introducing the animals of non-Cistercians into Cistercian flocks and herds, allowing these animals to have access to Cistercian breeding stock and pasturelands. These practices are first documented in 1228, but it is from about 1250 that charters record complaints about the practice of allowing "outsiders' animals" (*aliena animalia*) to pasture with Cistercian flocks.[101] It is likely that the problems

---

[97] *Cart. de Bonneval*, no. 161 (1244), describes passage free of tolls by Bonneval's animals through the Lozère. Bonneval also had a grange in that region at Anduze: Rodez, A. D. Aveyron, 3H series N, "Inv. de Bonneval," p. 52 (1247). Such activities in the Rhône basin are reflected also in *Statuta*, ed. Canivez, 1: 1200, no. 52, a dispute between Bonneval's monks and the men of the city of St. Gilles.

[98] Ibid., 1: (1214), no. 62.

[99] The tendency for mother-houses in this period to treat daughter-houses as appendages is suggested by the papal bull in the Bonnefont cartulary where the daughter-houses of Bonnefont are listed as properties in a papal confirmation of properties: *Rec. de Bonnefont*, no. 207 (ca. 1200).

[100] On the later commercialized traffic from the Quercy, see Bousquet, "Transhumance," 237–46; however, he ignores the ties between Bonneval and the Rhône valley, documented above.

[101] *Cart. de Léoncel*, no. 100 (1228), is an assurance that the donor will introduce only his own sheep and those of his men; ibid., no. 222 (1265) allows both Léoncel and the donor to introduce *aliena animalia*, but the donor's men may not do so; see also no. 252 (1284): "Quod omnes et singuli de ipsis monasterio et domo et ordine Lioncelli de cetero in perpetuum possint inducere animalia et greges suos animalium quorumcumque in dictorum territorium Castrinovi, . . . et omnes bestias suas et pastorum suorum, sive sint eorum proprie vel pro tempore, teneant eas ad medium crescementum."

arose among those who had earlier granted rights in the *estives* to the monks and now found that Cistercian commercialized use was excessive.[102] The contractual arrangements mentioned in the cartulary of Léoncel by which such animals were kept along with Cistercian flocks and herds were called *ad medium crescementum* or *a meis creis*.[103] These appear to have been analogous to the agricultural *complant* or *méplant* (*medium vestum*) agreements, or the later *gasaille* contracts.[104] If the analogy holds true, the contracts must have been designed to let outsiders entrust animals to Léoncel so that they could be taken to the highlands in the summer, provided with the salt necessary for health, and be bred, milked, and shorn. The term for the arrangement suggests that newborn lambs and other animals would have been divided between the monks and lay owners, but the terms of division may also have referred to the distribution of cheese and wool.

It was also for Provence that the first attempts to collect tithes on the transhumant flocks of the Cistercian houses of southern France are recorded. Such efforts by ecclesiastical corporations to tax these flocks by mid-thirteenth century are additional evidence of the extent of Cistercian pastoralism, and probably reflect the perception on the part of churchmen that the Cistercians owned and kept very large numbers of animals in the pasturelands of that region. As a result it became possible to deny Cistercian tithe exemption for that stock-raising in certain cases after 1215, despite the papal decisions in favor of the continued exemption of the food of Cistercian animals.[105] Evidence of this is found in a contract dated 1254, in which the churches of St. Félix of Valence and St. Peter of the Bourg of Valence agree in an arbitrated settlement on how to divide tithes which they both claimed on animals belonging to the Cistercians of Léoncel.[106] This contract makes no mention of any exemption for Léoncel. It is clear that Léoncel in this case is practicing transhumance, for animals born in territories belonging to one ecclesiastical corporation were moved, presumably for summer pasture, into territories under the control of the other house. Thus, the dispute had arisen because the two churches both were claiming tithes on Cistercian animals.

---

[102] Such a controvery arose with Léoncel in 1216: see ibid, no. 79 (1216).

[103] See the last document in n. 101.

[104] A similar term, *medium fenum*, seems to have concerned ownership of meadows (which the Cistercians had irrigated?): see *Cart. de Bonnevaux*, no. 267 (1164–1167), "Simon de Elbone dedit . . . fratribus Bonevallis . . . ad medium fenum in pratis de Aqua Bella, et habuit inde equum." On *medium vestum*, see Duby, *Rural Economy*, 139. "Contrats de gasaille," are found for Grandselve after 1250; see Mousnier, "L'abbaye," pp. 226–27.

[105] In the early exemptions, stock-raising was clearly tithe-exempt; after 1215 the situation seems to have caused dispute, but papal exemption generally continued. See discussion on tithe exemption, above, pp. 83–86, or the Bonnecombe papal confirmations, cited in chapter 4, notes 36–38 which clearly included "nutrimentum animalium." In the case of Léoncel, however, animals may have been considered subject to tithe because they belonged to others besides the monks—however, the documents are silent on whether disputes arose over this point.

[106] Cited in *Cart. de Léoncel*, no. 178 (1254). Printed in *Cartulaire de St. Pierre du Bourg-lès-Valence* no. 39, pp. 75–77.

Whatever the later controversy aroused by its aggressive stock-raising, pastoralism's large contribution to the success of early Cistercian houses in southern France cannot be overemphasized. In addition to providing individual abbeys of the order with significant amounts of cash which could be used for land purchase, the expanded pastoralism of the Cistercians in southern France had other effects. An increase in the proportion of livestock per acre on Cistercian holdings helped agriculture by improving land fertility. Increased numbers of animals also meant stronger plow teams, whether of horses or oxen, and greater possibilities for selective breeding. The sylvo-pastoral isolation which Cistercians as pastoralists shared with other new religious groups was probably one of the reasons why Cistercians came to be associated with deserted lands or forested wilderness, for medieval pastoralism was much more forest-oriented than pastoralism is today. The order's pastoralism was in many ways a last refuge from the growing urbanization of the twelfth century. At the same time, in seeking markets for its animals and animal products, the Cistercians developed significant ties to and commercial interests in regional cities. Perhaps even more than by its cereal cultivation and marketing of cereals, the increasingly commercialized pastoralism of the largest Cistercian houses in southern France made the Cistercians less rural, or at least less anti-urban in many respects, than their early propagandists would have us think. Pastoralism was the most solitary of all Cistercian economic activities, but at the same time it was also most responsive to demand from the growing cities of western Europe at the beginning of the thirteenth century.

VI. CISTERCIANS AND REGIONAL ECONOMIC GROWTH

From such evidence as has been presented here, what conclusions can be drawn about Cistercian agriculture? What general statements about the order's contribution to the economic development of southern France can be derived from a reassessment of the local archival evidence? It seems clear that although Cistercians in southern France can no longer be credited with extensive creation of new arable land through clearance and reclamation, they can be praised for their reorganization of lands by extensive purchases—a reorganization which created a firm economic base for the order's subsequent activities.

Although there are vague hints in the documents of anticipated or requested foundations which were never achieved, or daughter-houses founded only to be transformed into granges, the majority of Cistercian foundations and incorporations had achieved economic security by the mid-thirteenth century.[1] Of the forty-three monasteries considered here, all but two, la Gard-dieu and Ardorel, appear on the Cistercian tax rolls which began to be compiled in the middle of the thirteenth century (and, Sauvelade does not appear to have yet been part of the order), and only two other southern French houses, Léoncel and Franquevaux, both with known financial or administrative problems in the thirteenth century, were among the lowest contributors to general assessments. On the other hand, no southern French monastery is found among the order's most wealthy.[2] Generally, the successful Cistercian houses of the Midi appear to have been of only moderate size in terms of the order as a whole. It is these moderately sized houses, however, which are most characterized by the developments of new grange agriculture and intensified pastoralism discussed here, for they have provided the large majority of surviving records for this study. It was also the more successful Cistercian houses in the south which had the closest ties to the growing cities of the region.

The reorganization of land which Cistercians undertook in southern

---

[1] Expected expansion by Candeil, however, as suggested by the donation by Eleanor of Aquitaine of rights to found a religious house in honor of her son, did not occur: see Paris, B. N., Coll. Doat, vol. 115, fol. 168, cited in Rossignol, "Une charte d'Aliénor, duchesse d'Aquitaine de l'an 1172," *Revue d'Aquitaine et de Languedoc* 5 and 6 (1861), 224–28; the document is not dated and in my opinion is later than 1172.

[2] See Arne Odd Johnsen and Peter King, *The Tax Book of the Cistercian Order* (Oslo: Universitetsforlaget, 1979), 30–31 and Appendix A, Table 1, below, which extracts the southern French material from that assessment list. Locdieu's particular troubles are discussed in chapter V; those of Franquevaux are implied in Gorsse, *Valmagne*, 22–24.

France meant considerable upheaval for the local communities in which the order's houses were founded, but, even if traditional rural practices were disrupted and some existing inhabitants uprooted, Cistercian acquisition often gave many peasants new opportunities. In many cases, the disturbance to traditional economic patterns caused by Cistercian arrival in a neighborhood had a positive effect on all concerned. For lords, Cistercians provided cash to invest in new agricultural equipment or in foreign military adventures with potential returns such as the Crusades. For tenants, monks provided them an opportunity to free themselves from obligations on holdings which were fragmented and unproductive, and cash with which to consolidate other lands, to start new lives on the frontiers, or to move to the growing urban centers of the region.

Another immediate if indirect result of the foundation of Cistercian religious houses for the society into which the order came, and from which its monks and *conversi* were drawn, may have been a temporary acceleration of population expansion resulting from a shortening of time between generations. This shortening of the length of a generation would have occurred when the Cistercians provided a secure and respectable place of retirement for parents, since this allowed children to marry, to inherit, and to start families at a younger age. At least temporarily, population growth would have increased as a result of earlier retirement and earlier marriage. Thus, as Mundy has suggested with regard to urban charitable foundations, the establishment of new monasteries in the countryside, because they allowed a dignified retirement for the rural aged, spurred demographic growth.[3] New abbeys were also, of course, symptoms of growth.

Cistercians also contributed to the well-being of the rural communities into which they came by their reform ideals of asceticism and simplicity in religious services. The Cistercian return to a less complicated liturgy than that of the Cluniacs meant that the time devoted by the monks exclusively to prayer was reduced, that some part of each monk's daily routine was again devoted to economically productive manual labor in the fields.[4] By devoting time to both prayer and to his own support, a Cistercian monk was less economically dependent on the outside community than earlier monks who had devoted their lives wholly to prayer. Moreover, provisions in the order's statutes which forbade expensive colored glass and the use of precious metals for crucifixes and church plate were probably generally upheld although eventually Cistercian architecture at least became more elaborate.[5] The new Cistercian houses certainly did not replace the older,

---

[3] John H. Mundy, "Charity and Social Work in Toulouse, 1100–1250," *Traditio* 22 (1966): 203–87.

[4] Private communication, Br. Chrysogonus Waddell, Gethsemane Abbey.

[5] Surviving twelfth-century Cistercian churches demonstrate that the prohibition on extensive display even in building was upheld for some time: see Silvanès, Flaran, and Sénanque, as illustrated in Anselme Dimier, *L'Art cistercien; France*, 131–48, and 305–12. On churches, see Caroline A. Bruzelius, "The Twelfth-century Church at Ourscamp," *Speculum* 56 (1981): 28–40, and Marcel Aubert, *L'Architecture cistercien en France* (Paris: Editions d'art et d'histoire, 1943), passim.

more luxurious and more costly monastic centers, but they provided religious services for communities into which they came at considerably lower material cost; this is clear if the value of alms received and that of goods and services produced by monastic personnel themselves are taken into consideration.[6] Their abstemious ways had other economic consequences besides a cut-rate for religious services; in particular, resources not tied up in gold and silver ornaments could be reinvested in productive land. As Duby and Herlihy have both suggested for a slightly earlier period, medieval economic growth was often the result of putting treasure back into circulation, whereas stagnation resulted from hoarding it.[7] Thus, the Cistercians, by investing in land rather than freezing their wealth in church plate, were contributing to the amount of money in circulation and to the velocity of exchange. They were, therefore, in a small way responsible for the economic growth which is a normal response to such economic stimuli.[8]

After the mid-thirteenth century, Cistercians in southern France were also the founders of new towns. Military towns called *bastides* were established by some abbeys along the border with the English in Gascony.[9] These foundations which were made in concert with royal authorities reflect the increased political prestige of the order in the thirteenth century. Cistercians were well suited to such town-foundation, since the efforts associated with town organization and related recruiting activities were similar to those of the order's land acquisition. When *bastides* were successful, the conversion of Cistercian granges into towns made areas (where the monks had once replaced peasant populations with granges), again into centers of lay population. To inhabitants of the new *bastides* and others, moreover, Cistercians also provided new organizational ideas and patterns of agricultural management. Records of relatively early pasture disputes between the order and its neighbors, for instance, show that as the Cistercians became wealthier, their neighbors began to emulate their transhumance, and to contract with them for keeping animals. Lay populations may also have followed their lead in irrigation of meadows. It was probably such new monks who first introduced fulling mills into southern France; these new "industrial" facilities, like the many grinding mills maintained by the order, by harnessing available water power freed both Cistercian laborers and

---

[6] Twelfth-century attacks on Cluny were to some extent based on its wastefulness, particularly during the reign of Pons of Melgueil; Peter Damian in particular is known for his condemnation of Cluny for spending too much time in secular business, etc.: see Jean LeClercq, "The Monastic Crisis of the Eleventh and Twelfth Centuries," *Cluniac Monasticism in the Central Middle Ages*, ed. Noreen Hunt (Hamden, Ct.: Archon Books, 1971), 221.

[7] David J. Herlihy, "Treasure Hoards in the Italian Economy, 960–1139," *Economic History Review* 2nd ser. 10 (1957): 1–14, and Duby, *Early Growth*, 54–55.

[8] Harry A. Miskimin, *Money, Prices, and Foreign Exchange in Fourteenth-century France* (New Haven: Yale University Press, 1963).

[9] The Cistercian contribution to this growth was remarked by Marc Bloch, *French Rural Society*, 14–15, where he discusses founding of *bastides* by Grandselve. Unfortunately, he had mistaken *bastides* with new settlements like the eleventh-century *sauvetés*, such as those established by Moissac and St. Sernin as implied by documents found in Grandselve's cartularies, as discussed above, pp. 45–46.

other manpower for more productive activities.[10] Ultimately, it is within this larger context that the economic influence of the Cistercians in any region must be viewed.

In their economic activities Cistercians in southern France were also less isolated from growing urban centers than the traditional picture of Cistercian economies, and of the order's general aversion to towns, would have us believe. There has long been a tendency for economic historians to link Cistercian agricultural efforts in the countryside indirectly to the rebirth of towns and the growth of urban population.[11] In earlier models, however, Cistercian contributions were generally believed to have been indirect and inadvertent. Cistercians seeking rural solitude had accidentally created "bumper crops" which provided the food supplies for demographic growth and population migration to towns. A consideration of the southern French documents, however, shows that this explanation does not hold. These documents suggest, instead, that Cistercian ties to towns were actively sought, and that commercial privileges were valued because such towns provided markets for the order's surplus produce. Whatever the rhetoric of twelfth-century Cistercians, the order's return to the primitive Benedictine tradition was not a turning away from growing urban centers; the urban ties established by the order's monasteries were not accidental.[12]

Although both cereal production on large-scale granges and transhumant pastoralism continued to dominate southern French Cistercian economic practice, from at least the beginning of the thirteenth century, if not before, the order did not confine its activities solely to the countryside. Particularly in their emphasis on pastoralism, Cistercians were quite responsive to the economic demands of neighboring towns.[13] In particular, the order was not content simply to sell its produce in those towns, but if possible, to sell that produce with preferred status. The archives show that Cistercian abbeys sought numerous commercial rights and properties in those centers most conveniently located for their trade. Commercial rights or preferences were gained through exemptions from tolls and taxes allowing the order to move animals and goods toll-free to urban centers and once there to buy and sell without paying market taxes; urban activities also included acquisition of urban mills, gardens, houses, workshops, and hospices for their members.

---

[10] Bautier, "Les plus anciens mentions," 568.

[11] As discussed in chapter 1, pp. 7–8.

[12] On Cistercian anti-urban polemic, see Lester K. Little, *Religious Poverty and the Profit Economy in Medieval Europe* (Ithaca, N.Y.: Cornell University Press, 1978), 27–41.

[13] The notion that Cistercians were really much more town-oriented than our traditional image of them makes sense; in coming to an accommodation with towns they foreshadow the mendicant orders of the thirteenth century, playing a transitional role between traditional, rural monasticism (if such a thing existed) and the urban monasticism of the friars. Evidence for such a conclusion about the order is briefly cited in the following notes, but the urban activities of the Cistercians in this region are beyond the scope of this study. See Berman, "Monastic Hospices," forthcoming.

Interestingly enough, it was in the granting of such exemptions rather than in donations of land *per se* that the "great" families of the Midi most patronized the Cistercians, as was discussed in relation to Valmagne's acquisitions in Languedoc. Both Valmagne and Silvanès were given commercial rights in many of the towns of that region and together they shared control of a hospice in Montpellier which was used by their monks and *conversi* when travelling to the city. Valmagne also kept boats in the village-port of Mèze for fishing and for transporting goods to markets along the lagoons of the Mediterranean coast.[14] Grandselve's rights on the Garonne have also been mentioned, but the monks of that abbey did not limit their commerce to Toulouse and the regions north and west of that city which were accessible by that river. With its daughter-houses of Candeil and Calers, Grandselve also acquired extensive rights to move its produce by way of Carcassonne to Narbonne and the vicinity of its daughter-abbey Fontfroide, and the Mediterranean coast.[15] So did the monks of Boulbonne.[16]

Like Grandselve, Fontfroide acquired extensive concessions of rights to pass free of tolls particularly on the coastal plain with its goods including rights to take its animals into Catalonia.[17] Fontfroide, like other Cistercian houses of the region, undoubtedly took its wool and cheese, its animal hides, and its cereals to nearby commercial centers on the Mediterranean coast, where important woolen and leather industries were beginning to develop at this time.[18] Other abbeys had similar commercial ties to cities with growing industrial and commercial sectors. Although there is no explicit evidence of toll exemptions or of acquisition of urban property in the charters of Berdoues and Gimont, the records of the General Chapter suggest that those abbeys concentrated commercial interest on Toulouse.[19] This was probably the case for most of the Cistercian houses of Gascony. The abbey of Bonneval in the Rouergue, like Valmagne and other Cistercian houses on the Languedoc coast or those in Provence, aimed its commercial activities toward the cities and ports of the Rhône, taking its animals into the marshes of the Camargue for wintering and selling its produce in markets along the coastal lagoons, as is implied by reports of conflict between Bonneval and the merchants of St. Gilles.[20] Bonneval, however, also had

---

[14] Montpellier, A. D. Hérault, film, "Cart. de Valmagne," vol. I, fols. 132v ff. (1148), and *Histoire Général de Languedoc*, 5: col. 1097, no. 571: 3.

[15] Conveyances of exemption from passage tolls for Grandselve's animals and its trade are included in documents in Paris, B. N., Coll. Doat, vol. 76, nos. 102 (1187), 74 (1203), 56 (1198), 83 (1208), 21 (1189), etc.

[16] Paris, B. N., Coll. Doat, vol. 83, Boulbonne, nos. 91 (1217), 96 (1188), 101 (1192), 136 (1203).

[17] Grèzes-Rueff, "Fontfroide," 277.

[18] On woolen textiles in Languedoc, see Bautier, *Economic Development*, 208, and Reyerson, "Le rôle de Montpellier dans le commerce des draps de laine avant 1350," *Annales du Midi* 94 (1982): 17–40.

[19] *Statuta*, ed. Canivez, 1: 1200, no. 69.

[20] Ibid., 1: 1200, no. 52.

houses and commercial rights in the regional commercial center of Rodez, where it may have provided many animal products to the developing local woolen industry.[21] Like Bonneval's, the monks of Bonnecombe had rights in Rodez as well as storage facilities at Lodève for goods *en route* to Languedoc.[22] But, the stress of Bonnecombe's commercialism generally was toward the region west of Rodez, in the direction of Albi, Gaillac, and presumably Toulouse, rather than toward Languedoc.[23] This difference in commercial ties between the two neighboring abbeys, Bonneval and Bonnecombe, is presumably a reflection of filiation. Bonnecombe was linked with its mother-abbey Candeil to the Clarevallian house of Grandselve and to Toulouse; Bonneval had links to Mazan and les Chambons in the Cévennes and from there to Provence.

Acquisition of workshops and houses in the cities and towns in which the order marketed its goods and animals was followed by mills in those same centers by Bonnecombe, its mother-house Candeil, and by Grandselve.[24] Urban mills in towns such as Toulouse and Agen, or in the vicinity of Castres, indicate that those houses were interested in the industrial processes tied to animal husbandry, particularly fulling and tanning.[25] In these cities the order also began to develop commercial facilities designed to house the personnel or handle the goods which Cistercians were offering for sale; several of these hospices would later become Cistercian colleges.[26] For instance, in Montpellier, Silvanès shared control with Valmagne of an urban hospice founded in the 1160s, administered by its lay donor, with a separate endowment for its needs, where the monks and *conversi* of Silvanès and Valmagne could stay when travelling to Montpellier to transact Cistercian business.[27] Like that in Montpellier, a hospice was founded in Nîmes in 1215 under the control of the monks of Franquevaux, but was clearly open to use by all members of the order.[28] Other hospices were founded by Grandselve and Boulbonne in Toulouse.[29] These urban holdings

---

[21] *Cart. de Bonneval*, nos. 4 (1163), 12 (1169), 24 (1176), 61 (1184), 85 (1192), 161 (1244), 70 (1185), 80 (1190), 81 (1190), 125 (1202), and 126 (1202).

[22] Rodez, Soc. des lettres, Bonnecombe, "Cart. de Moncan," nos. 165 (1246) and 166 (1245); "Cart. d'Iz et Bougaunes," nos. 29 (1179), 180 (1203), and 211 (1215); and *Cart. de Bonnecombe*, nos. 190 (1216), 226 (1215), and 296 A (1217). Bonnecombe may also have had a hospice in Rodez; see ibid., no. 286 A (1205).

[23] Rodez, A. D. Aveyron, H Bonnecombe, "Cart. de Bernac," nos. 117 (1220), 130 (1253), 119 (1228), 167 (1252), and 173 (1253).

[24] Paris, B. N., Coll. Doat, Candeil, vol. 114, nos. 55bis (1265) and 112 (1192). Toulouse, A. D. Haute-Garonne, 2 H Grandselve, vol. 77, no. 273 (1249); Paris, B. N., Coll. Doat, Grandselve, vol. 77, no. 102 (1224); and Paris, B. N., Paris, Grandselve, Latin MS 9994, fol. 283v; also a reference to a "stare ad lavandam lanam vostram."

[25] Paris, B. N., Coll. Doat. vol. 114, Candeil, no. 55 (1208), vol. 77, Grandselve, no. 102 (1224), *Layettes*, vol. II, no. 3539 (1246), and III, 3860 (1250); for other parts of Europe see Lekai, *Cistercians*, 321–23.

[26] Louis J. Lekai, "Le collège St. Bernard au moyen âge," *Annales du Midi* 85 (1973): 251–60, discusses that of Grandselve in Toulouse.

[27] *Cart. de Silvanès*, nos. 462 (1161) and 463 (1161), Montpellier, A. D. Hérault, film "Cart. de Valmagne," vol. 1, fols. 146v–47v (1161), and Berman, "Monastic Hospices," forthcoming.

[28] Nîmes, A. D. Gard, Franquevaux, H 74 (1215).

[29] Lekai, "Collège," 251–60.

are perhaps the best evidence of an increasingly commercial focus in Cistercian economic activity by the late twelfth century. The location and the separate endowment of these hospices made them easily convertible into Cistercian colleges. In their original guise, however, the hospices, more than any other Cistercian properties, reflect the order's stress on establishing ties between its rural activities, like agriculture and pastoralism, and the urban markets which they supplied. This was a Cistercian economic innovation. Earlier Benedictine communities had tended to allow markets to develop on their doorsteps; Cistercians in southern France, instead, emphasized and promoted economic ties to towns at some distance, although always within easy reach, of an abbey's seclusion.

Many commercial privileges came to the Cistercians in the first decades of the thirteenth century, at a moment when the order's monks were acting as preachers against heresy and as leaders of the Albigensian Crusade.[30] Not surprisingly, Cistercians benefited economically in the role as arbiters of orthodoxy, receiving a surge of donations particularly from the rich and powerful in the years just before the Crusade.[31] Indeed, after the Crusade began, northern conquerors and their southern rivals vied with each other in conveying gifts of rents and exemption from passage tolls to the order, such as the annual 1800 *solidi* from the tolls of Béziers, which Simon de Montfort granted to Fontfroide in 1217.[32] In the treaty of 1229 for which the abbot of Grandselve was a negotiator for the peace between northern Crusaders and the defeated southerners, Count Raymond of Toulouse promised large sums to a number of Cistercian houses, and agreed to rebuild the church of Candeil which had been destroyed during the war.[33] The Cistercians also gained from their role as defenders of orthodoxy by the acquisition of the property of a number of condemned heretics.[34]

Urban ties, political ties, the beginnings of commercial interests, all are manifestations of a new non-rural outlook among the Cistercians in the Midi. This adaptability of Cistercian ideals to local conditions in that region demonstrates that the order's economic "principles" were probably never as rigid or invariable as popular belief has made them. Although certain activities of Cistercian monks in this region may appear to contradict the order's early ideals, the issues on which the order was flexible had little to do with its spiritual life, and should not be used as indices of religious

---

[30] Preaching against heresy in the south was begun by Bernard of Clairvaux and continued by later Cistercians: see Angleton, "Cistercian Missions," pp. 120 ff.; Moore, "St. Bernard's Mission," 1–10; Christine Thouzellier, *Hérésie et hérétiques: Vaudois, Cathares, Patarins, Albigeois* (Rome: Edizioni di Storia e Letteratura, 1969), chap. 6; for a general view of the crusade see Wakefield, *Heresy*, 96–129.

[31] See, for example, the numbers of donations to Berdoues and Valmagne during the years immediately before 1210.

[32] Carcassonne, A. D. Aude, MS H 211, "Inv. de Fonfroide," fol. 16 (1217), a gift by Simon de Montfort. Southerners would necessarily have to counter or regrant such bequests; see gifts by Raymond of Toulouse, et al. to Valmagne: Montpellier, A. D. Hérault, film "Cart. de Valmagne," vol. I. fols. 150r–58v.

[33] Devic and Vaissete, *Histoire générale de Languedoc*, 8: col. 886.

[34] Grèzes-Rueff, "Fontfroide," 268–70, and Carcassonne, A. D. Aude, H 211, MS, "Inv. de Fonfroide," fol. 79.

decadence. Probably in every region where the order made foundations or incorporated existing religious settlements its economic ideals had to adapt to local conditions. This is particularly likely to have been so once the order had developed beyond the heroic eremitical stage. In southern France such accommodation to local conditions was a necessity because there was little wilderness available and because most clearance and reclamation had already been completed before Cistercians entered the region. There, much more than in Burgundy, the dream of monastic settlement in untouched places was nothing but an unworkable ideal. Probably everywhere in Europe, however, from at least the 1150s on, foundation of the order's abbeys and granges in untouched and inaccessible places became unlikely, since the order itself was not likely to accept new houses or found them in areas where their chances for success were low. Once the new monks were joined by *conversi* in their economic activities, once abbeys began to recruit able, farm laborers from the vicinity, once adequate resources were required before a community could be established, the ideal of the heroic foundation in the desert became no more than a memory or an illusion, one no longer compatible with other parts of the Cistercian economic program.[35]

The question of personnel and their recruitment is significant in this respect. As has been argued, in the twelfth century *conversi* could only have been recruited as part of the larger process of land acquisition: that is, in already-cultivated agricultural lands. Even in areas of settled agriculture, it was eventually the lack of sufficient numbers of *conversi*, a situation which coincided with the end of land acquisition in the mid-thirteenth century, which brought Cistercian agriculture into difficulties. Even in the earliest years, however, there is a possibility that the number of *conversi* was the limiting factor in some Cistercian rural activities, and that it was as a result of too few laborers that Cistercians often did not undertake labor-intensive activities like viticulture or clearance. Throughout the documents there are numerous indications that Cistercians seemed to prefer the managerial tasks of repurchasing and consolidation over hard physical labor, but the lack of sufficient laborers may well have been the decisive factor that prevented Cistercians from undertaking whatever clearance and reclamation may still have been possible in southern France. In any case, surviving documents show that whether it was drainage in the Camargue or clearance of woods near Villelongue, such work was undertaken by peasants rather than by Cistercians.[36]

---

[35] This in itself is probably another argument for why the early legislative ideals of the order must be viewed as a cumulative collection of statutes rather than as the workings of a single, inspired brain. A careful consideration of the economic viability under twelfth-century conditions of such an economic plan would undoubtedly show that certain parts would simply never fit together if rigidly upheld.

[36] See Villelongue's removal of peasants in the 1160s to a *villeneuve* above, pp. 47–48, or the concern that peasant settlement might increase near Lodève, *Cart. de Silvanès*, no. 381 (1147), or assarts made by tenants removed from Grandselve's land, B. N. Latin MS 11008, Grandselve, no. 178 (11-78), or p. 45, n. 64 above.

The charters and cartularies which they drew up are firm evidence that Cistercian monks were good managers and that the order attracted its membership from a particularly able segment of society. In offering a religious life to knights and peasants, the Cistercian order attracted individuals who had been good administrators as laymen and who would take vital roles in the new monastic administration. These men were careful record-keepers, organizers, and farmers of the large tracts of agricultural land and pasture which the order acquired. They knew the importance of exemption from tithes and other taxes, and they took great care to insure the future prosperity of their monastic houses. There is little evidence that such groups had been brought up to the rough, physical labor of clearance and other reclamation, nor would it have made sense for them to have undertaken such back-breaking and often poorly repaid efforts unless they were absolutely necessary, since start-up costs were high and areas available for reclamation poor.

Southern French Cistercian houses did have sufficient monks and *conversi* to allow the introduction of cultivation on large, compact granges, and to undertake the intensified pastoralism which characterizes the order's economies in that region. That such agriculture and pastoralism were possible is primarily a result of the order's early acquisition policies. Existing settlement patterns in southern France meant that a considerable reorganization of fragmented holdings into large granges had to be undertaken before those holdings became the profitable agricultural units which could practice Cistercian grange agriculture. Once such granges were introduced, the order could profit from efficiencies in scale, the lower dependency costs of *conversi*, and the tax savings of exemptions, as well as reap all the benefits of the increased yields resulting from capital investment, better agricultural techniques, and other improvements. In pastoralism, the numbers of animals were increased under Cistercian management by the introduction of transhumance. The increased numbers of animals owned by the order provided not only the funding for land acquisition, but contributed to increased land fertility. Finally, not only did the production of wool, leather, parchment, and other animal products provide for the internal needs of Cistercian communities, but the sale of surplus animal products was especially profitable because of growing urban demand for animal products as both food and industrial raw materials. Such consequences of Cistercian economic practice, whether or not the order's new managers were conscious of them, provided individual houses with an increasing number of profitable granges and contributed greatly to the order's wealth during its first century of existence.

In addition to being successful at reorganizing land, the Cistercians were also good farmers. They were successful as farmers because agriculture and pastoralism had become part of their daily life—part of their labor of prayer. Thus, in twelfth-century southern France these monks continued to be closer to the working and management of the soil than their Benedictine predecessors. The monks' and lay brothers' enthusiasm and zeal for their chores, coupled with an intensive knowledge of local agricultural conditions, were vital to the economic success of many of the order's southern French houses. These were men who knew the land intimately because they had been born there; they were concerned with agricultural cultivation and land management because such activities were part of their praise of God. They also undoubtedly worked more effectively on that land than their peasant predecessors, because they were well fed. Despite all this, they were neither pioneers nor frontiersmen; the sites on which they chose to live were neither deserts, wildernesses, nor the "horrible places" of the Old Testament, except in their imaginations. They did not choose the unrewarding hard work of clearance when it was possible to do otherwise.

If this is the case, then it might be asked why Cistercians did not revise their program to fit local practice more closely. Surviving records for the twelfth-century Cistercian General Chapter are probably too fragmentary to provide an answer on that level, although recorded legislation does suggest that the chapter was not unresponsive to criticism about the order's "excessive expansion" or "greed for land." Those abbots did make gestures toward halting or limiting expansion in response to such criticism.[37] On a local level, the only southern French evidence which might bear directly on the problem of the order's early divergence from its ideals is the account of the foundation of Silvanès written in the 1160s. The chronicle of Silvanès was designed to inspire, to influence donors to Silvanès, and to aid that abbey in competition for endowment with its Cistercian neighbors of Valmagne and Bonnecombe, but Hugh Francigena, the monk who wrote it, may have inadvertently provided answers to the question of why the ideals were not revised.[38] In many ways his account epitomizes the idealization of early Cistercian foundations and economies; he claims to have talked with those who had participated in the abbey's foundation, but his picture of the creation of the abbey is more literary than real.[39] He describes a wilderness site although there is clear evidence to the contrary in the charters of land acquisition.[40] Despite considerable local evidence that the monks

---

[37] *Statuta*, ed. Canviez, 1: 1152, no. 1; 1180, no. 1, etc.

[38] *Cart. de Silvanès*, intro. by Verlaguet, discusses its authorship, p. xii, n. 2.

[39] Loc. cit., 372, "precipiente hoc michi domino Pontio, abbate meo, et singula suggerente que vel ipse vidit ab initio vel audivit ab eis qui viderunt et interfuerunt, imo ab ipsis qui primi institutores et fundatores hujus loci fuerunt,—nonnullis etiam de patribus nostris hoc attestantibus qui solo auditu cuncta didicerant, sed illorum testimonio innititur magis nostra narratio qui ab initio cuncta noverunt et laboris et patientie priorum participes extiterunt, videlicet Hugonis, presbiteri, et Raimundi Alzarani, de quorum testimonii veritate nullus permittitur dubitare."

[40] Loc. cit., 381-82, "venerunt in terram, que Camares nuncupatur, que terra nemorosa est et silvis obsita, ardua montibus, devexa collibus, irrigua fontibus, rivis atque fluminibus, . . . locum, qui antiquitus Silvanium a silvis dicebatur, . . . in quo casulas propriis manibus fabricantes manserunt, bestiis sociati, quotidiano tamen labori insistentes, dumeta falcibus resecantes, terram ligonibus proscindentes, locum habitabilem ex inhabitabili reddiderunt." Clearly, the literary aspects of this text deserve more study.

of Silvanès had been preceded at their site by peasant inhabitants, Hugh did not acknowledge, or perhaps did not recognize, that Silvanes's foundation took place in anything but the prescribed way. There is no suggestion of apology here; Hugh makes no attempt to excuse a foundation contrary to ideals. In fact, he appears to have seen no discrepancy between his account and reality. If this is true on a local level, it seems probable that the twelfth-century Cistercian abbots in the General Chapter, similarly, would have seen no significant difference between their ideals and practice anywhere in western Europe. A completely "other-worldly" view of the situation, that such details were not important, may have taken over. These are the kinds of details about which Bernard of Clairvaux, for instance, was famous for being vague. Another view, suggested by Lekai, is that in the struggle to found and maintain new religious houses, such minor factors as a variation from statute or "legislative suggestion" in land-holding acquisitions, seemed unimportant.[41] It is nonetheless difficult to reconcile such vagueness with the managerial abilities suggested by the order's surviving archival records.

Perhaps more pertinent is the fact that although not in total conformity to all the stated ideals of the order, variations from ideals in southern France were limited. Moreover, Cistercian abbeys in southern France did create something new, something immediately recognizable as different from the economic practice of earlier monks. They established large agricultural granges, employed *conversi* and hired laborers in place of dependent peasants, and exploited the possibilities of an intensified pastoralism, although they rarely cut down trees or uprooted brush in order to create new arable land. When they obtained already-existing villages, or churches and tithes, or other manorial properties, such acquisitions were a step toward transforming those manorial resources into new non-manorial land holdings—Cistercian granges—rather than the beginnings of seigneurialism. A transformation of the rural world did take place under the impact of Cistercian settlement, but it was not the transformation from forest or desert to quiet abbey space envisioned in idealized accounts. Instead, existing villages and peasant settlements were reorganized into a new kind of property—the grange.

The new Cistercian agriculture on such granges made the order's houses in southern France increasingly prosperous. In addition, by the beginning of the thirteenth century the growing political power of the order in the region aided it in the acquisition of wealth, although not necessarily adding to the esteem in which the order was held. Many individual Cistercians, either promoted to bishoprics or to abbacies of the order's most powerful houses, rose to positions of political power in the thirteenth century. Such power and prestige, of course, added to the ability of the Cistercian order in the Midi to acquire and protect monastic property, but the growing politicization of the order made its efforts to preach against heresy on the

---

[41] Lekai, "Ideal," 13 ff.; see also Auberger, cited above p. 33 n. 20.

eve of the Albigensian Crusade ineffective.[42] Unlike the new mendicant preachers, the white monks could no longer approach the towns as exemplars of monastic poverty. By 1200, the order's monks entered the towns of the south expecting both privileged status in commercial dealings and treatment as distinguished visitors. This power of the order within the Church is much more significant as a mark of the growing distance from early Cistercian ideals than any aspect of the order's economic practice. Indirectly, of course, the eventual loss of religious idealism may have been a result of rural economic success.

Cistercian houses were founded in southern France, however, primarily because they could provide a new ascetic presence for the region. Although it was for religious purposes that the Cistercian order had been introduced, by reorganizing fragmented, worn-out land holdings into granges, esteeming manual labor, practicing capable management, and introducing labor-saving devices and better agricultural techniques into southern France, it also made a contribution to the economic prosperity of the region. Monks and *conversi* did transform the land and managers did establish a new kind of agricultural regime with strong ties to transhumant pastoralism and urban markets. The order's economic program as described here, based on Cistercian granges, Cistercian *conversi*, Cistercian pastoralism, Cistercian tithe exemption, and finally, on Cistercian commercial privileges and links to urban areas, was considerably different in its details from the economic model which has traditionally been put forth.

The older model of Cistercian success was based on a "frontier" thesis which cannot be applied to the order's foundations in southern France. By assuming that the benefits of frontier farming applied to Cistercian economic practice, earlier historians limited their search for explanations of the growth of Cistercian wealth. Although mentioning granges and *conversi* and tithes, they did not consider far enough the economic consequences of these Cistercian innovations. By looking at the Cistercian economy from a new point of view, one which dismisses the narrative accounts as generally idealizing and propagandistic, this reassessment has confirmed the linkages between many parts of the traditional picture of Cistercian practice, but has shown that granges and tithes and *conversi* were tied to one another in a different way than has previously been thought. In so doing, this study does not deny that Cistercians transformed the southern French countryside through their creation of granges. It does attempt to explain better what that transformation entailed and how the monks contributed to the feeding of growing cities in southern France by the late twelfth century, and to what extent the rural economic life of Cistercian monasteries had already begun to be tied to the re-emerging markets and urban centers of southern Europe by the middle of the thirteenth century.

---

[42] Arnaud Amalric, is most noteworthy: see Foreville, "Arnaud," passim. There were others, like Foulque, bishop of Toulouse, mentioned by Foreville, but see also Mundy, *Liberty*, 78–85, Wakefield, *Heresy*, 83–91, and Thouzellier, *Hérésie*, chap. 6.

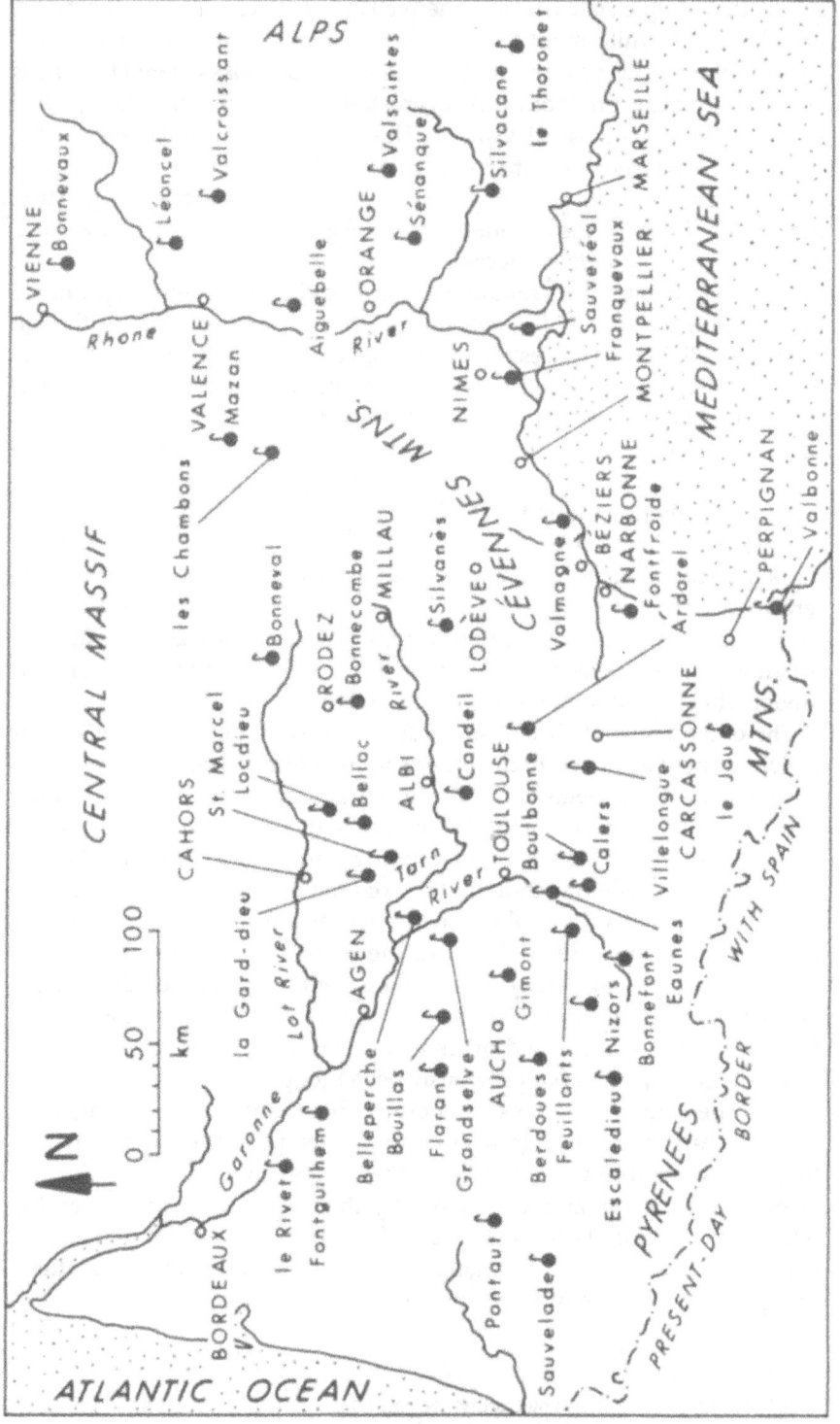

Map 1: Cistercian Monasteries in Southern France

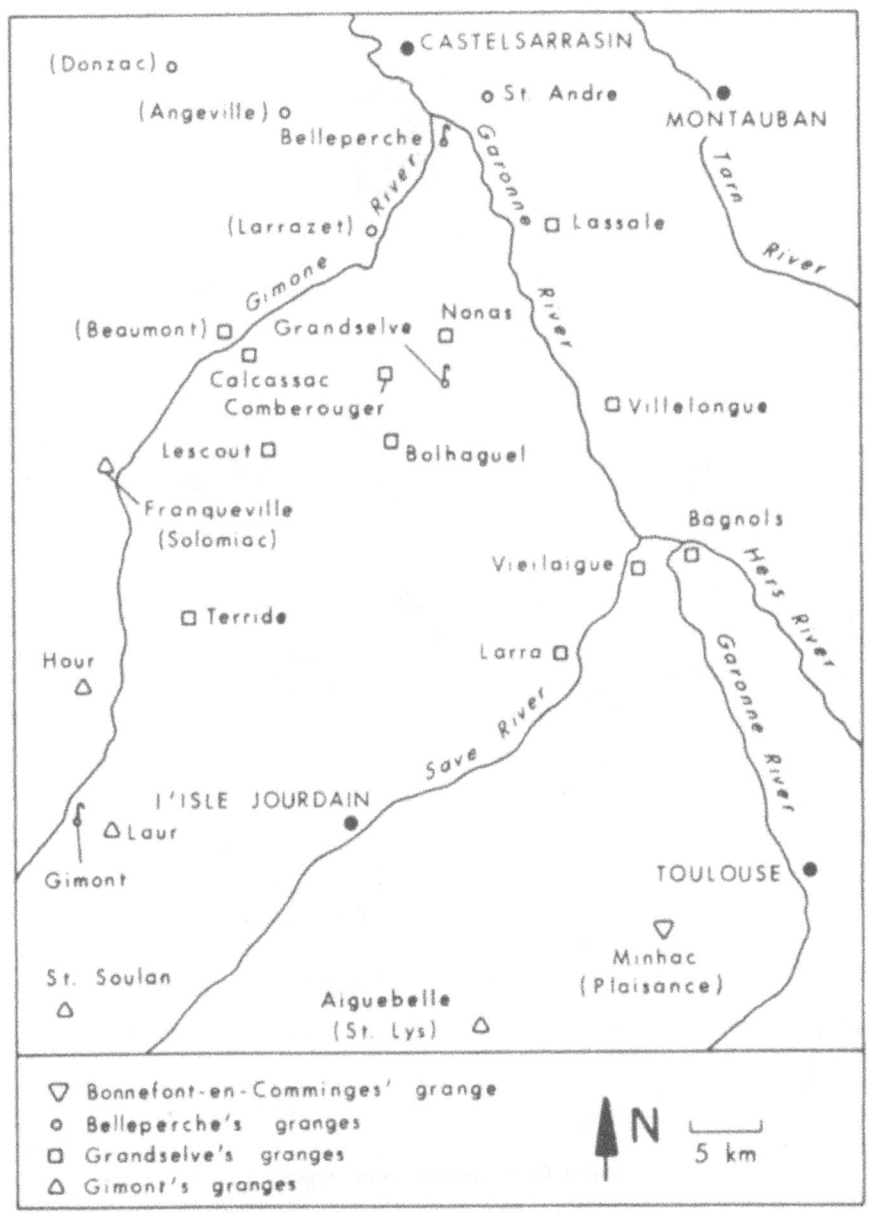

Map 2: Cistercian Granges near Toulouse

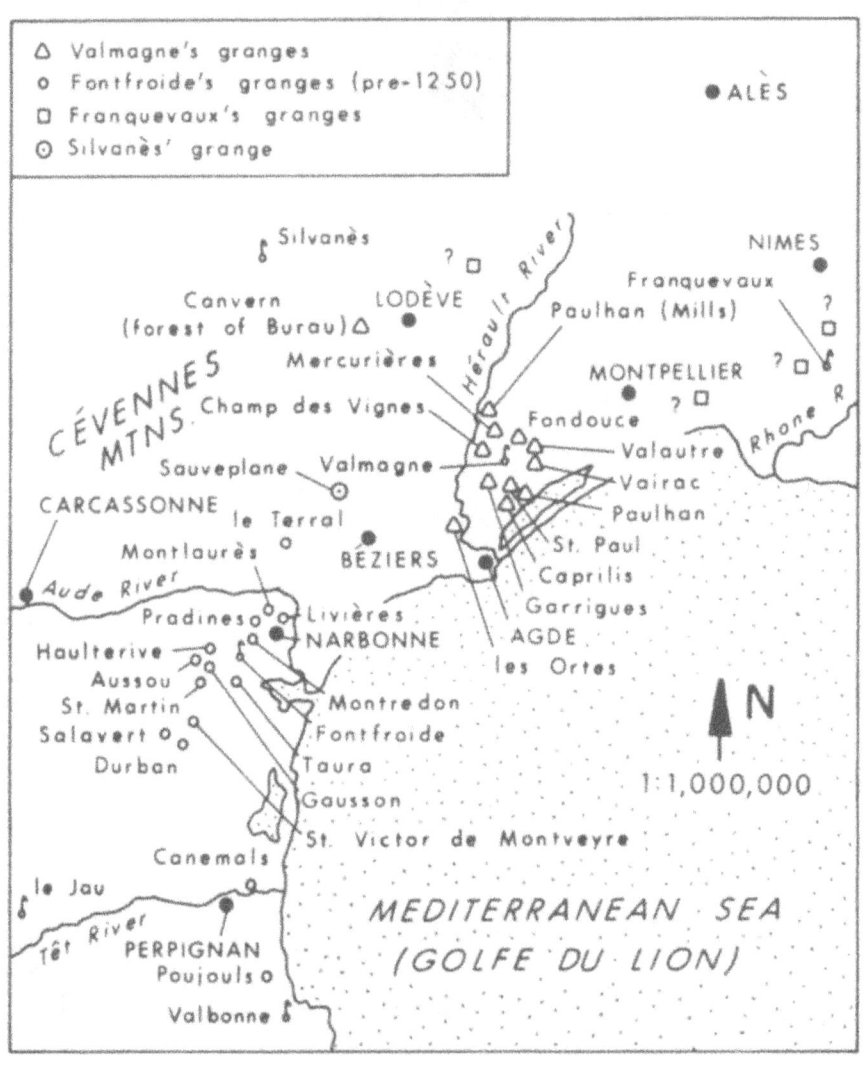

Map 3: Cistercian Granges in Languedoc

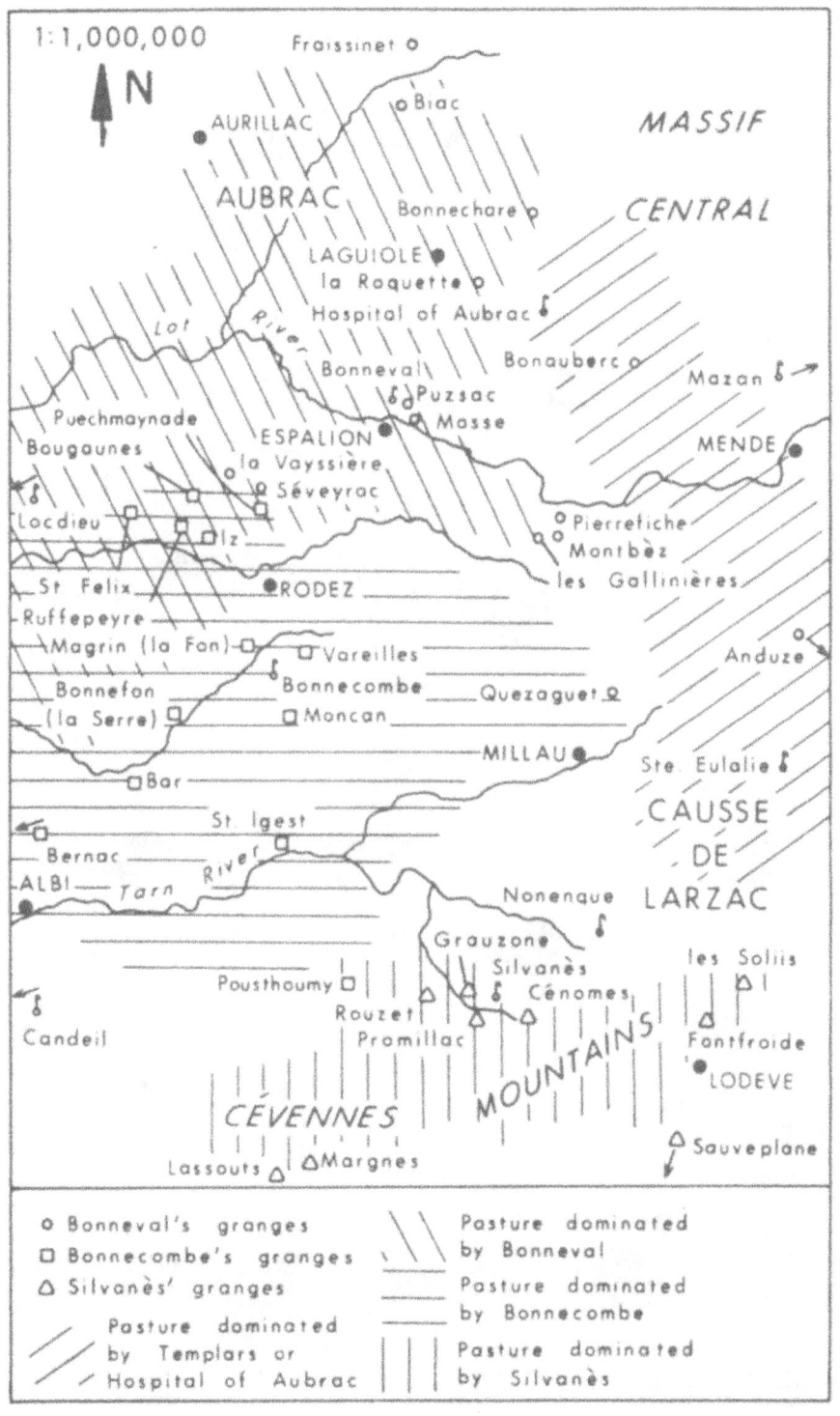

Map 4: Cistercian Granges and Pastoralism in the Rouergue

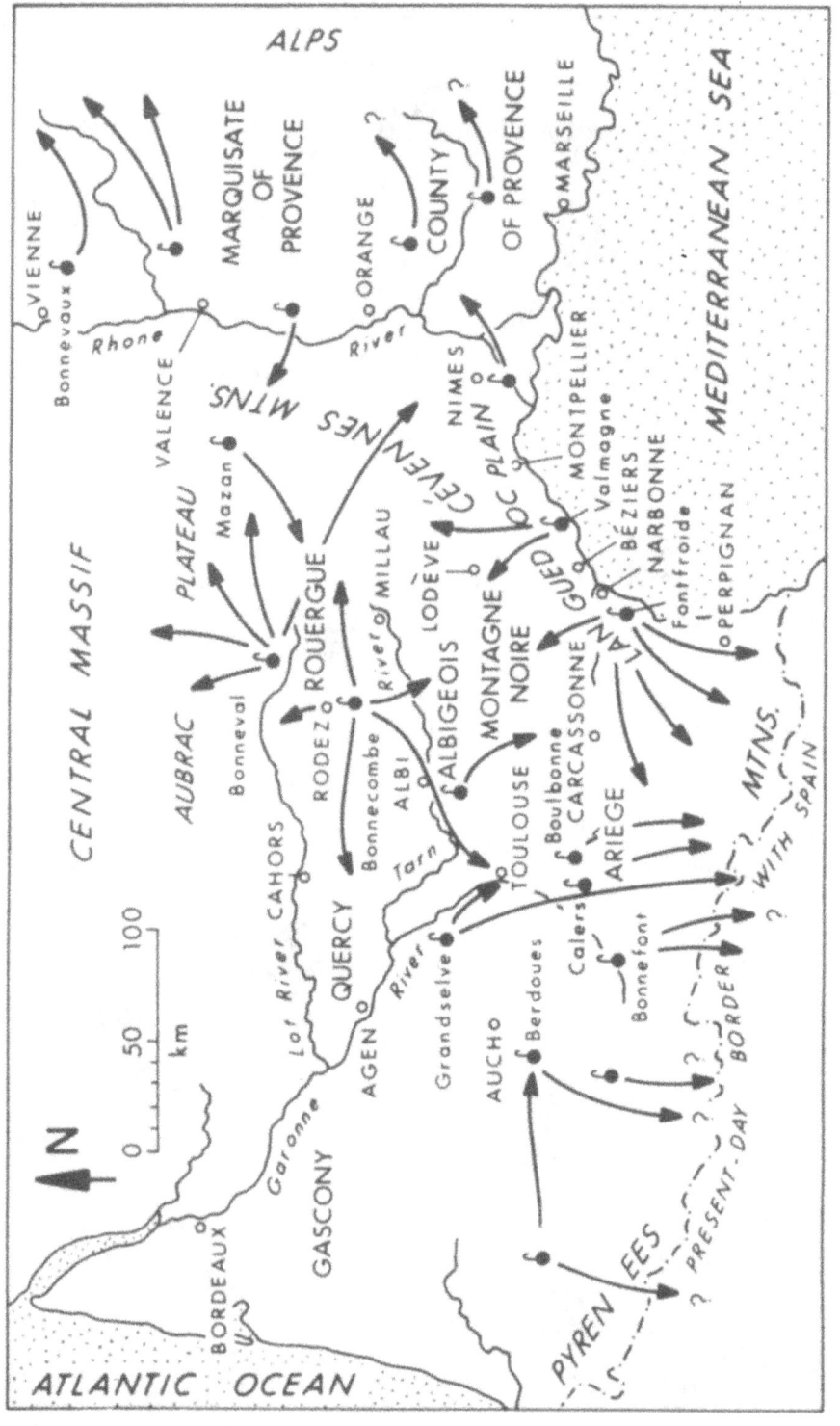

Map 5: General Directions of Cistercian Transhumance

TABLE 1

CISTERCIAN FOUNDATIONS AND INCORPORATIONS IN SOUTHERN FRANCE WITH DATE, FILIATION, MOTHER-ABBEY, AND DIOCESE[a]

Abbey Name	Filiation[b]/ Mother-abbey	Foundation (f)/ Incorporated (i)	Earlier group Known?	Diocese
Aiguebelle	M./ Morimond	f. 1137	no	Valence
Ardorel	P./ Cadouin	f. 1110s/ i. 1147	hermits G. de S.[c]	Albi
Belleperche	Cl./ Clairvaux	f. 1110s/ i. ca. 1145	hermits G. de S.	Toulouse
Belloc	Cl./ Clairvaux	f. 1133–44/ i. 1144	hermits?	Rodez
Berdoues	M./ Morimond	f. 1137?/ i. 1142	earlier monks?	Auch
Bonnecombe	Cl./ Candeil	f. 1162–63	no	Rodez
Bonnefont	M./ Morimond	f. 1122/ i. 1136–39	Hospitallers	Toulouse
Bonneval	Cî./ Mazan	f. 1147	no	Rodez
Bonnevaux	Cî./ Cîteaux	f. 1119	no	Vienne
Bouillas[d]	M./ Escaledieu	f. 1150	no	Auch
Boulbonne	M./ Bonnefont	f. ca. 1150–63	hermits	Toulouse
Calers	Cl./ Grandselve	f. 1148	no	Toulouse
Candeil	Cl./ Grandselve	f. 1152	no	Albi
Chambons (les)	Cî./ Sénanque[e]	f. 1152	no	Vivarais
Eaunes	M./ Berdoues	f. 1120/ i. 1150	hermits G. de S.	Toulouse
Escaledieu (l')	M./ Morimond	f. ?/ i. 1137–42	hermits	Toulouse
Feuillants	P./ Locdieu	f. ?/ i. 1162	hermits? G. de S.[f]	Toulouse
Flaran	M./ Berdoues[g]	f. 1151	no	Auch
Fontfroide	Cl./ Grandselve	f. ca. 1093/ i. 1145[h]	Benedictines	Narbonne
Fontguilhem	P./ Gondon	f. ca. 1124/ i. 1147	hermits G. de S.	Bordeaux
Franquevaux	M./ Morimond	f. 1143	no	Nîmes
Gard-dieu (la)	Cî./ Obazine	f. 1140s?/ i. 1150	hermits[i]	Cahors
Gimont	M./ Berdoues	f. 1147	no	Auch
Grandselve	Cl./ Clairvaux	f. 1114?[j]/ i. 1145	hermits G. de S.[k]	Toulouse
Jau (du)[l]	P./ Ardorel	f. 1147?/ i. 1162	Benedictine	Elne
Léoncel	Cî./ Bonnevaux	f. 1137	no	Valence
Locdieu	P./ Dalon	f. 1124/ i. 1162	hermits-Dalon	Rodez
Mazan	Cî./ Bonnevaux	f. 1120/ i. ca. 1136	hermits?	Vivarais
Nizors[m]	M./ Bonnefont	f. 1184	no	Toulouse
Pontaut	P./ Dalon	f. 1115–25/ i. ca. 1150	hermits-Dalon	Adour
Rivet (le)	P./ Pontaut	f. ?/ i. 1147?	hermits-Dalon	Bordeaux
St. Marcel	P./ Cadouin	f. 1160s/ i. 1175	hermits	Toulouse
Sauvelade	M./ Gimont	f. 1127/ i. 1287	Benedictine	Lescour
Sauveréal[n]	Cî./ Bonnevaux	f. 1173	??	Arles
Sénanque	Cî./ Mazan	f. 1148	no.	Avignon
Silvacane	M./ Morimond	f. 11th C./ i. 1147	St. Victor/hermits?	Aix
Silvanès	Cî./ Mazan	f. 1132/ i. 1138	hermits	Rodez

TABLE 1 (*Continued*)

Abbey Name	Filiation[b]/ Mother-abbey		Foundation/ Incorporated	Earlier group Known?	Diocese
Thoronet (le)	Cî./	Mazan	f. 1136	no?	Fréjus
Valbonne	Cl./	Fontfroide	f. 1242	no	Elne
Valcroissant	Cî./	Bonnevaux	f. 1188	no	Valence
Valmagne	Cî./	Bonnevaux	f. 1138/ i. 1155	Ardorel hermits	Agde
Valsainte	M./	Silvacane	f. 1188	no	Apt
Villelongue[o]	M./	Bonnefont	f. 1140s/ i. 1150–51	hermits	Carcassonne

[a] Information drawn primarily from Janauschek, *Originum*, passim, but see also Berman, "Growth," forthcoming.

[b] P. = Pontigny, Cî = Cîteaux, Cl. = Clairvaux, M. = Morimond.

[c] Gerald of Salles, western French eremitical founder and reformer and follower of Robert of Arbrissel.

[d] Also known as Portaglonium.

[e] After 1214, Bonneval.

[f] Independent house under Locdieu earlier.

[g] Or l'Escaledieu.

[h] Became daughter-house of Grandselve in 1142.

[i] Attached to nuns of Fontmourle who were attached to Obazine in 1140s.

[j] By 1130 there were coenobitic monks there.

[k] Possibly a connection with Robert of Arbrissel, the neighboring nuns of Lespinasse, and Fontévraut; there was an altar dedicated to Mary Magdalen by 1117 at Grandselve.

[l] Also known as Clariana.

[m] Also called la Bénisson-Dieu.

[n] Originally called Ulmet, moved to Silva Albaronis in 1194 and to Silva Réal in 1243. Land at Ulmet had originally been held by the nuns of St. Cesarius of Arles and it is possible that there were earlier attempts at founding a religious settlement there.

[o] Called Campagnes until move to new site ca. 1170 at Villelongue.

TABLE 2

SURVIVING DOCUMENTS AND TOTAL CASH EXPENDITURES FOR CISTERCIAN ABBEYS

Documents counted as Donations, Sales and Other Onerous Contracts[a], or Mortgages

Abbey[b]	Number of Surviving Documents				Total Cash Expenditures in *Solidi*[c]
	Donations +	Sales +	Mortgages =	Total	
Aiguebelle[d]	5	5		10	—[e]
Ardorel	9	5		14	—
Belleperche	24	10		34	—
Belloc	10	10	3	23	—
Berdoues					
1130–1149	2	1	1	4	37.0
1150–1169	22	41	27	90	4,239.0
1170–1189	32	36	32	100	6,688.5
1190–1209	59	44	18	121	2,520.0
1210–1229	64	58	31	153	4,995.5
1230–1249	59	30	6	95	3,103.5
undated	59	90	85	234	
Totals	297	300	200	797	21,583.5
Bonnecombe					
1150–1169	10	6		16	227.0
1170–1189	125	152	2	279	11,872.0
1190–1209	99	222	3	324	29,054.5
1210–1229	82	139		221	38,406.0
1230–1249	78	97	6	181	56,022.0
undated	27	18	1	46	
Totals	421	634	12	1067	135,581.5
Bonnefont	37	16	3	56	—
Bonneval[f]					
1150–1169	12	8		20	4,938.0
1170–1189	24	31		55	7,649.0
1190–1209	16	15		31	2,167.0
1210–1229	7	3		10	1,000.0
1230–1249	11	3		14	450.0
undated	9	16		25	
Totals	79	76		155	16,204.0

TABLE 2 (*Continued*)

Abbey[b]	Number of Surviving Documents				Total Cash Expenditures in *Solidi*[c]
	Donations +	Sales +	Mortgages =	Total	
Bonnevaux					
1110–1129	12	3		15	—
1130–1149	8	7		15	1,557.0
1150–1169	28	32		60	11,532.0
1170–1189	70	32	1	103	5,794.0
1190–1209	52	13	1	66	1,418.0
1210–1229	3			3	—
1230–1249	1			1	220.0
undated	156	155		311	19,074.0
Totals	330	242	2	574	39,595.0
Boulbonne					
1150–1169	17	5		22	1,240.0[g]
1170–1189	23	16	2	41	1,942.5
1190–1209	25	30		55	6,834.0
1210–1229	16	11		27	248.0
1230–1249	7	5		12	185.0
undated	10			10	—
Totals	98	67	2	167	10,449.5
Calers	39	21		60	
Candeil					
1150–1169	33	7		40	269.0
1170–1189	21	83	1	105	12,361.0
1190–1209	20	41	1	62	6,299.0
1210–1229	12	6		18	364.0
1230–1249	11	11		22	2,225.0
undated	10	7		17	—
Totals	107	155	2	264	21,518.0
Franquevaux					
1130–1149	1			1	—
1150–1169	22	17		39	7,795.0
1170–1189	78	96		174	26,091.0
1190–1209	18	33		51	29,075.0
1210–1229	8	4		12	840.0
1230–1249	6	4		10	1,680.0
undated	8	5		13	10,750.0
Totals	141	159		300	76,231.0

TABLE 2 (*Continued*)

Abbey[b]	Number of Surviving Documents				Total Cash Expenditures in *Solidi*[c]
	Donations +	Sales +	Mortgages =	Total	
Gimont					
1130–1149	6	13		19	250.0
1150–1169	97	182	3	282	4,653.0
1170–1189	200	148	1	349	3,396.0
1190–1209	52	30	1	83	274.0
1210–1229	11			11	—
1230–1249	4	1		5	—
undated	17	10		27	
Totals	387	384	5	776	8,573.0
Grandselve					
1130–1149	12	8	1	21	96.0[g]
1150–1169	269	210	7	486	2,497.5
1170–1189	373	311	14	698	12,427.5
1190–1209	135	42	1	178	745.0
1210–1229	44	12	1	57	5,450.0
1230–1249	50	2		52	—
undated	35	13	1	49	5.0
Totals	918	598	25	1541	21,221.0
Léoncel					
1130–1149	1			1	—
1150–1169	6	5	1	12	1,800.0
1170–1189	6	8		14	3,668.0
1190–1209	16	9		25	6,278.0
1210–1229	17	16		33	2,610.0
1230–1249	25	27		52	2,400.0
undated	6			6	5,883.0
Totals	77	65	1	143	22,639.0
Locdieu	31	19		50	—
Nizors	87	21	1	109	—
Silvanès[h]					
1130–1149	54	60		114	3,500.0
1150–1169	82	230		312	21,678.0
1170–1189	7	19		26	4,785.5
1190–1209	1	3		4	110.0
1210–1229	3	2		5	60.0
undated	11	11		22	
Totals	158	325		483	30,133.5

TABLE 2 (*Continued*)

Abbey[b]	Number of Surviving Documents				Total Cash Expenditures in *Solidi*[c]
	Donations +	Sales +	Mortgages =	Total	
Valmagne					
1130–1149	9	11		20	3,146.5
1150–1169	32	95	3	130	6,568.5
1170–1189	79	171	1	251	45,687.5
1190–1209	77	151		228	20,866.5
1210–1229	8	3	1	12	1,000.0
undated	12	6	0	18	174.0
Totals	217	437	5	659	77,443.0
Villelongue	34	29		63	—
Grand Totals	3,506	3,578	261	7,345	481,172.0

[a] Since these can include promises of admission into an abbey as brethren (certainly a potentially onerous undertaking), exchanges, payments in kind, and payment of annual rent, that are not included in the cash expenditures column, there may be a slight discrepancy between numbers of onerous contracts and amounts spent.

[b] Fontfroide is not included in this table which does not generally rely on inventories for its data; those for Fontfoide, however, do give some interesting data, which has been compiled by Grèzes-Rueff, "Fontfroide," 263, 265, 266. Unfortunately, it is not possible to interpolate all the data with any accuracy from his graphs. Some highs and lows are clear and can be cited: in the years 1210–14 alone, 25,100 sol. were expended by Fontfroide in land acquisition; in the period 1215–19, however, only 310 sol. were expended. The highest five-year interval was 1250–54 in which 95,400 sol. were expended by the monks of Fontfroide. Numbers of purchases in the intervals 1145–65 were approximately 19; from 1165–85: 165; from 1185–1205: 91; from 1205–25: 50; from 1225–45: 39.

[c] These are in the coinages of the region: deniers of Morlaas, Melgueil, Cahors, Rodez, etc.; all have roughly equivalent values; but see notes below.

[d] Where very few documents have survived, the numbers have been given as totals only for the period through 1249.

[e] Where numbers of surviving documents are small, such amounts become increasingly less representative; they have therefore not been included.

[f] References in the later inventory (3H Nouv. ser. 19 "Inv. de Bonneval") are not included; the statistics include only documents printed in the abbey's cartulary. These are but a limited number of what must have originally existed, since a fire destroyed most of the abbey's archives.

[g] Amounts expressed in *solidi* of Toulouse (a double coin) in the period up to 1210 have been converted to the more standard *solidi* of Melgueil.

[h] The actual Silvanès cartulary breaks off rather abruptly; later charters included were collected by Verlaguet in his edition of the medieval codex.

TABLE 3

NUMBER OF DOCUMENTED GRANGES FOR CISTERCIAN HOUSES IN SOUTHERN FRANCE

Abbey	Filiation	Number of Granges	Source of Information	Date
Aiguebelle	M.	1	Chartes d'Aiguebelle, no. 40	1207
Belleperche	Cl.	6	Doat, vols. 91–92	to 1250
Berdoues	M.	8	Cart. de Berdoues	dated 1257–58+
Bonnecombe	Cl.	9	Bonnecombe cartularies	dated 1246–54+
		5	Cart. de Bonnecombe, no. 64	1178
		7	Ibid., no. 72	1220
		8	Ibid., no. 81	1244
Bonnefont	M.	5	Rec. de Bonnefont, no. 78	1165 conf.
		7	Ibid., no. 340	1246 conf.
Bonneval	Cî	6	Cart. de Bonneval, no. 3	1162
		11	Ibid., no. 70	1185
Boulbonne	M.	4	Doat, vol. 83, fol. 147	1183
		5	Ibid., fol. 189	1186
		5	Ibid., fol. 309	1188
		8	Doat, vols. 83–86	to 1250
Candeil	Cl.	5	Albi, "Inv.," no. 146	1230–32
Fontfroide	Cl.	5	Carcassonne, "Inv.," fol. 2	1162
		14	Loc. cit.	ca. 1200
Franquevaux	M.	5	Nîmes, H43, no. 2	1173
Gimont	M.	5	Cart. de Gimont,	dated ca. 1215+
Grandselve	Cl.	5	Gallia Christiana, 13, col. 18	1142
		13	Toulouse, 108H56, fol. 2v.	1161
		11	Ibid., no. 112	1212
Léoncel	Cî	5	Cart. de Léoncel, no. 13	1165
		7	Ibid., no. 25	1176
Locdieu	P.	6	Doc. de Locdieu, no. 44	1182–85
Silvanès	Cî	9	Cart. de Silvanès	dated ca. 1160+
		7	Ibid., no. 2	1154
		9	Ibid., no. 1	1162
Valmagne	Cî	1	"Cart. de Valmagne," I, fol. 6r	1162
		8	Ibid., I, fol. 6r ff.	1185
		12	Ibid., passim	ca. 1210

APPENDIX I

ASSESSMENTS OF SOUTHERN FRENCH CISTERCIAN ABBEYS BY FILIATION[a]

From *Secundum Registrum*[b] dated ca. 1330[c] using assessment called *Contributa Duplex*,[d] and From *Report of the Commission of Regulars* 1768,[e] by descending order of Assessment in the Former

Abbey	*Contributa Duplex* (ca. 1330)	1768 Value	1768 Members
Filiation of Cîteaux			
Valmagne (p. 40)	48 L. 10 sol.	9,000 L.	8 monks.
Thoronet[f] (le) (p. 40)	34 L. 10 sol.	8,900 L.	8 monks.
Bonneval[g] (p. 40)	30 L. 15 sol.	20,000 L.	22 monks.
Sauvéréal[h] (p. 40)	30 L.	n/a	n/a[i]
Chambons (p. 40)	29 L. 5 sol.	13,000 L.	9 monks.
Silvanès (p. 40)	18 L.	5,200 L.	6 monks.
Mazan (p. 40)	18 L.	9,200 L.	11 monks.
Bonnevaux[j] (p. 38)	17 L. 16 sol. 6 den.	7,300 L.	7 monks.
Sénanque (p. 40)	15 L. 16 sol. 6 den.	4,000 L.	4 monks.
Valcroissant (p. 40)	13 L. 10 sol.	1,800 L.	2 monks.
Léoncel (p. 40)	4 L. 10 sol.	7,000 L.	5 monks.
Gard-dieu (la)	n/a	3,600 L.	3 monks.
Compare Cîteaux	n/a	70,000 L.	60 monks.
Filiation of Clairvaux			
Grandselve (p. 54)	75 L. 11 sol. 6 den.	25,225 L.	15 monks.
Fontfroide (p. 76)	67 L.	14,300 L.	10 monks.
Belleperche (p. 54)	60 L.	14,417 L.	11 monks.
Bonnecombe (p. 76)	60 L.	12,000 L.	12 monks.
Candeil (p. 76)	30 L.	7,538 L.	5 monks.
Belloc (p. 56)	18 L.	3,485 L.	3 monks.
Calers (p. 76)	16 L. 10 sol.	4,024 L.	8 monks.
Valbonne (p. 76)	11 L. 5 sol.	n/a	n/a
Compare Clairvaux	n/a	78,711 L	54 monks.
Filiation of Morimond			
Boulbonne (p. 78)	30 L.	17,562 L.	13 monks.
Bouillas[k] (p. 88)	27 L.	5,800 L.	9 monks.
Berdoues (p. 80)	16 L. 10 sol.	5,997 L.	6 monks.
Gimont (p. 88)	16 L.	10,400 L.	9 monks.
Bonnefont (p. 80)	12 L.	7,400 L.	8 monks.
Escaledieu (p. 80)	12 L.	9,000 L.	7 monks.
Aiguebelle (p. 90)	12 L.	1,756 L.	2 monks.
Eaunes (p. 88)[l] (p. 90)	10 L. 10 sol.	3,848 L.	4 monks.
Nizors (p. 88)	10 L.	5,200 L.	4 monks.
Feuillants (p. 84)	9 L.	n/a	n/a
Flaran (p. 88)	7 L.	3,927 L.	4 monks.
Villelongue (p. 88)	7 L.	3,983 L.	2 monks.

APPENDIX I (Continued)

Abbey	Contributa Duplex (ca. 1330)	1768 Value	1768 Members
Filiation of Citeaux			
Silvacane (p. 90)	6 L. 12 sol. 6 den.	n/a	n/a
Valsainte (p. 90)	5 L. 5 sol.	2,441 L.	3 monks.
Franquevaux (p. 80)	4 L. 10 sol.	5,109 L.	10 monks.
Sauvelade[m]	n/a	2,699 L.	2 monks.
Compare Morimond	n/a	20,800 L.	30 monks.
Filiation of Pontigny			
Pontaut (p. 48)	48 L.	3,257 L.	4 monks.
le Rivet (p. 50)	30 L.	5,099 L.	3 monks.
Locdieu[n] (p. 48)	18 L. 15 sol.	5,000 L.	5 monks.
Ardorel (p. 48)	18 L.	5,000 L.	6 monks.
Fontguilhem (p. 48)	15 L.	4,175 L.	2 monks.
St. Marcel (p. 48)	15 L.	2,250 L.	3 monks.
le Jau[o] (p. 48)	12 L.	n/a	n/a
Compare Pontigny	n/a	26,831 L.	25 monks.

[a] This is done by filiation, since comparisons between houses within a single filiation are more accurate than those which go across filiation lines because these early assessments were done within the filiation. Mother-houses from Burgundy are included in the 1768 survey for the sake of comparison; they do not seem to have been included in the earlier assessments.

[b] Published as *The Tax Book of the Cistercian Order*, ed. Arne Odd Johnson and Peter King (Oslo: Universitetsforlaget, 1979). Page number references in the table are to this edition, rather than to the pages of the manuscript.

[c] This date is established ibid., 27–28.

[d] By using the double assessment rather than the single one, amounts are more easily compared.

[e] Numbers are drawn from the table entitled "Personnel and Financial Status of Cistercian Monasteries in France in 1768 according to the report of the 'Commission of Regulars,'" Louis J. Lekai, *The White Monks* (Okauchee, Wisc.: Our Lady of Spring Bank Press, 1953), 277–83.

[f] Also called Floregia.

[g] Identified as Ruthenensis.

[h] Also called Ulmet.

[i] No assessment given.

[j] Identified as in the diocese of Vienne.

[k] Also called Portaglonium.

[l] There is also a reference in *Secundum Registrum* to Alius in Vasconia, with *Contributa Duplex* given as 13 L.; this is possibly a misreading of Alnis in Vasconia, a name for Eaunes, in which case, Eaunes has been misidentified by the editors.

[m] It is possible that the omission of Sauvelade from the *Secundum Registrum* lists provides a terminus ante quem for the initial compilations of the order's taxes: Sauvelade was only incorporated by the order in 1287.

[n] Described as in Ruthenensis.

[o] Also known as Clariana.

APPENDIX II

PART ONE

DISTRIBUTION OF PROFITS FROM AGRICULTURE GIVEN DIFFERENT ESTIMATES OF RENTS, YIELDS, TITHES, AND LABOR COST

This APPENDIX is intended to show that the available information about medieval yields, rents, etc. can give a coherent picture of who gets the harvest using the equation cited in the text of chapter 4:

GROSS HARVEST
= RENT + SEED + TITHE + LABOR COST + PEASANT'S PROFIT.

The tables below demonstrate that there were inherent limits to the system of portioning out the medieval harvest, dictated primarily by low yields in the Middle Ages, for if the landlord and church took too much, the peasant starved. If the peasant got too much he would either himself become a lord, or his landlord would institute new customs to reap some of the benefits.

Yield/Seed	Rent + Tithe	Seed[a]	Peasant's Share[b]	Comment
\multicolumn{5}{l}{Case One: Rent is Fifteenth Percent, Tithe is a Tenth. Depending on Yields,[c] 100 Bu. Gross Yield Will Be Distributed as Follows:}
1.5/1	15 + 10	67	8	Less than subsistence?
2.0/1	15 + 10	50	25	Adequate year?
2.5/1	15 + 10	40	35	Profitable year?
3.0/1	15 + 10	33	42	Too much compared to lord.[d]
3.5/1	15 + 10	29	46	Too much compared to lord.
\multicolumn{5}{l}{Case Two: Rent is a Fifth, Tithe is a Tenth. Depending on Yields, 100 Bu. Gross Yield Will Be Distributed as Follows:}
1.5/1	20 + 10	67	3	Less than subsistence?
2.0/1	20 + 10	50	20	Bare subsistence?
2.5/1	20 + 10	40	30	Adequate year?
3.0/1	20 + 10	33	37	Profitable year?
3.5/1	20 + 10	29	41	Too much compared to lord.
\multicolumn{5}{l}{Case Three: Rent is a Quarter, Tithe is a Tenth. Depending on Yields, 100 Bu. Gross Yield Will Be Distributed as Follows:}
1.5/1	25 + 10	67	⟨2⟩	Impossible to survive.
2.0/1	25 + 10	50	15	Bare subsistence?
2.5/1	25 + 10	40	25	Adequate year?
3.0/1	25 + 10	33	32	Profitable year?
3.5/1	25 + 10	29	36	Profitable year?

Yield/Seed	Rent + Tithe	Seed[b]	Peasant's Share[c]	Comment
Case Four: Rent is Thirty Percent, Tithe is a Tenth. Depending on Yields, 100 Bu. Gross Yield Will Be Distributed as Follows:				
1.5/1	30 + 10	67	⟨7⟩	Impossible to survive.
2.0/1	30 + 10	50	10	Less than subsistence?
2.5/1	30 + 10	40	20	Bare subsistence?
3.0/1	30 + 10	33	27	Adequate year?
3.5/1	30 + 10	29	31	Profitable year?

[a] SEED equals 1/Yield times 100 bu. in this case.
[b] PEASANT'S SHARE equals LABOR COST (the amount necessary to feed the peasant and his family annually for that amount of land in proportion to his total acreage) + PEASANT'S PROFIT.
[c] For justification for these amounts, see chapter 4 notes.
[d] Rent too low.

Cases Three and Four above show that if yields were too low, they placed a ceiling on rents, which could not exceed peasant subsistence for long. Only if yields were relatively high, could rents exceed twenty or twenty-five percent. Since we know that yields above 3.5 were very unusual and that the range between 1.5 and 2.5 was often normal, it must have been possible for peasants to survive if they got between 20 and 35 bushels of every 100 in their gross harvest. There is little to suggest that peasants normally made a profit on their agriculture during this period, therefore, this amount—called PEASANT'S SHARE—in these cases, must be approximately equal to the amount necessary to feed the peasant and his family.

PART TWO

EFFECTS WHEN CISTERCIANS WORKED LAND BY LAY-BROTHERS AND IT WAS TITHE EXEMPT

A: EFFECTS OF TITHE EXEMPTION ALONE

The tithe, a ten percent levy on gross profits, represents an especially significant percentage of net harvest after seed, if yields are very low, but becoming less significant as yields increase.

Yield/Seed	Seed	Net Harvest After Seed	Tithe	(Tithe/100-Seed) Increase
2.0/1	50	50	10	.20
2.5/1	40	60	10	.17
3.0/1	33	67	10	.15
3.5/1	29	71	10	.14

In comparison to the earlier lord, the tithe-exempt Cistercian abbey on the same land had a relative increase in rent or profit, simply on account of the tithe exemption, without taking any labor savings into account and without any improvement in productivity. This relative increase varies with the percentage of gross harvest attributed to rent under prior ownership, but is particularly high when rents are low, a result of yields being low.

Rent	Rent + Tithe	Rent + Tithe/Rent Minus 1 = Tithe/Rent = Increased Profit with Exemption		
15	25	10/15	=	.67
20	30	10/20	=	.50
25	35	10/25	=	.40
30	40	10/30	=	.33

B: EFFECTS OF *CONVERSI* LABOR

Without any improvements of scale, technology, or capital investment, the total cost of labor on Cistercian-owned land would be less than on the same land previously cultivated by peasants, because those tenants had higher dependency costs for labor. Because *conversi* did not incur the cost of feeding pregnant women, children, and elderly (remember, the revenues for the last for the monastery itself would have come from rent as landlord), the amount of the harvest consumed as food for laborers would be significantly lower. The problem, of course, is how much lower.

Using the initial equation above, it is possible to calculate a percentage increase or improvement in Cistercian rent or profit over that of the previous lord or owner. It would be equal to A (the percentage of harvest saved in labor food cost) times LABOR plus RENT divided by RENT:

$$\text{INCREASE WITH } CONVERSI = (((A*LABOR) + RENT)/RENT) - 1$$

$$= (A*LABOR)/RENT.$$

When the savings for tithe exemption is also included, the equation is:

$$\text{INCREASE WITH } CONVERSI \text{ AND TITHE EXEMPTION} = (((A*LABOR)$$

$$+ RENT + TITHE)/RENT) - 1 = ((A*LABOR) + TITHE)/RENT.$$

Although we do not know the exact amount of A, nor the exact amount the peasant comfortably consumed, a variety of estimates can be tabulated which give the percentage improvements with Cistercian *conversi* labor and tithe exemption, as follows:

APPENDIX II

		Percentage Increases			W. *Conversi* (A∗Labor)/Rent where A is			*Conversi* and Exemption ((A∗Labor) + Tithe)/Rent where A is		
		Equation used: Percentage savings:								
	Y/S	Rent + Tithe	Seed	P's Labor	30%	40%	50%	30%	40%	50%
Rent is 15%	1.5/1	15 + 10	67	8	<u>16</u>	21	27	<u>83</u>	88	93
	2.0/1	15 + 10	50	25	<u>50</u>	<u>67</u>	83	<u>117</u>	<u>133</u>	150
	2.5/1	15 + 10	40	35	70	<u>93</u>	117	137	<u>160</u>	183
	3.0/1	15 + 10	33	42	84	<u>112</u>	<u>140</u>	151	<u>179</u>	<u>207</u>
	3.5/1	15 + 10	29	46	92	123	<u>153</u>	159	189	<u>220</u>
Rent is 20%	1.5/1	20 + 10	67	3	<u>5</u>	6	8	<u>55</u>	56	58
	2.0/1	20 + 10	50	20	<u>30</u>	<u>40</u>	50	<u>80</u>	<u>90</u>	100
	2.5/1	20 + 10	40	30	45	<u>60</u>	75	95	<u>110</u>	125
	3.0/1	20 + 10	33	37	56	<u>74</u>	<u>93</u>	106	<u>124</u>	<u>143</u>
	3.5/1	20 + 10	29	41	62	82	<u>103</u>	112	132	<u>153</u>
Rent is 25%	1.5/1	25 + 10	67	⟨2⟩	—	—	—	—	—	—
	2.0/1	25 + 10	50	15	<u>18</u>	<u>24</u>	30	<u>58</u>	<u>64</u>	70
	2.5/1	25 + 10	40	25	30	<u>40</u>	50	70	<u>80</u>	90
	3.0/1	25 + 10	33	32	38	<u>56</u>	<u>64</u>	78	<u>91</u>	<u>104</u>
	3.5/1	25 + 10	29	36	43	58	<u>72</u>	83	98	<u>112</u>
Rent is 30%	1.5/1	30 + 10	67	⟨7⟩	—	—	—	—	—	—
	2.0/1	30 + 10	50	10	<u>10</u>	<u>13</u>	17	<u>43</u>	<u>47</u>	50
	2.5/1	30 + 10	40	20	20	<u>27</u>	33	53	<u>60</u>	67
	3.0/1	30 + 10	33	27	27	<u>36</u>	<u>45</u>	60	<u>69</u>	<u>78</u>
	3.5/1	30 + 10	29	31	31	41	<u>52</u>	64	75	<u>85</u>

N.B. When PEASANT'S labor portion is small there will be a tendency for the 30% columns to be more reliable; when PEASANT'S labor portion is large the 50% columns will be more reliable. In each of the two right hand sections there will be a tendency for the most reliable numbers (underlined) to follow the diagonal from top left to bottom right under each subsection.

BIBLIOGRAPHY

Unpublished Archival Sources in France

ALBI: A.D. Tarn,
  H1-9 Ardorel,
  H38 "Répertoire de Candeil: 1739-1741,"
  I/J74/3, Candeil.
CARCASSONNE: A.D. Aude,
  H211, "Inventaire de la manse conventuelle de Fonfroide."
MONTAUBAN: A.D. Tarn-et-Garonne,
  H1 Belloc,
  H3 Belleperche,
  H59 Grandselve.
MONTPELLIER: A.D. Hérault,
  film no. 1 Mi 260-261 (private deposit), "Cartulaire de Valmagne," 2 vols.
  H series, "Villemagne copy" of Valmagne cartulary.
MONTPELLIER: Société Archéologique de Montpellier,
  "Cartulaire dit de Trencavel (XIe-XIIIe)."
NARBONNE: Bibliothèque Municipale de Narbonne,
  MS 259, "Inventaire de la manse abbatiale de Fonfroide,"
  MS 260 "Cauvet Notes,"
  MS "Inventaire de l'archévêché de Narbonne."
NIMES: A.D. Gard,
  H33 "Inventaire ou cartulaire de Franquevaux."
  H34-H103, Franquevaux, liasses of documents.
PARIS: Archives Nationales,
  L1009 bis, Grandselve.
PARIS: Bibliothèque de l'Arsenal,
  MS 6470, Belloc.
PARIS: Bibliothèque Nationale:
  Collection Doat,
    vol. 59, Fontfroide;
    vol. 70, Villelongue;
    vols. 76-80, Grandselve;
    vols. 83-86, Boulbonne;
    vols. 91-92, Belleperche;
    vols. 114-115, Candeil;
    vols. 138-139, Bonnecombe;
    vol. 150, Silvanès;
  Collection Duchesne, vol. 118;
  Latin MSS 9994, 11008, 11009, 11010, and 11011, Grandselve;
  Latin MS 10975, Locdieu;
  Latin MS 11012 Eaunes;
  Latin MS N.A. 1698 Belloc.
  Collection des Benedictins: 12751, 12752, 12756.
RODEZ: A.D. Aveyron,
  2H non coté, Bonnecombe, "Cartulaire de Bernac,"
  Anc. coté 267-268, Bonnecombe, "Inventaire de Serres,"
  2H liasses 9-88, Bonnecombe, "Registre de Pousthoumy" (anc. coté 198),
  3H nouv. ser. 19, Bonneval "Inventaire."
RODEZ: Société de lettres, sciences et arts de l'Aveyron,
  MSS 1, 2, 3, Bonnecombe:
    "Cartulaire de Magrin,"

"Cartulaire de Moncan,"
"Cartulaire d'Iz et Bougaunes."
TOULOUSE: A.D. Haute-Garonne,
  108H Boulbonne, liasse and registres 1–3,
  H Calers, liasses 4–20,
  H Eaunes, liasse no. 5,
  108H Grandselve, liasses nos. 1–60,
  H Nizors, liasses 1–21 and "Inventaire de Nizors,"
  7D Collège de Saint Bernard, (Grandselve), nos. 37, 41, 138, 141,
  Ordre de Malte, Ste. Eulalie (Locdieu documents).
TOULOUSE: Archives Municipales,
  MS 342, Belleperche.
TOULOUSE: Bibliothèque Municipale
  MS 152 (Grandselve)
  MS 638 (Boulbonne).

PUBLISHED DOCUMENTS AND TEXTS

"L'abbaye d'Eaunes," ed. D. Garrigues. *Bulletin de la société archéologique du Midi* (Toulouse) 22 (1953): 78 ff.
Barrière-Flavy, Casimir. *L'abbaye de Calers*. Toulouse: Chauvin, 1887–89.
———. *L'abbaye de Vajal dans l'ancien comté de Foix (preuves)*. Toulouse: n. p., n. d.
Bernard of Clairvaux. *De Consideratione*, trans. and ed. John D. Anderson and Elizabeth T. Kennan. Kalamazoo, Mich.: Cistercian Publications, 1976.
Brunel, Clovis, ed. *Les plus anciennes chartes en langue provençale*. Paris: Picard, 1926, and supplement, Picard, 1952.
Cartulaire et archives de l'ancien diocèse et de l'arrondissement administratif de Carcassonne, ed. M. Mahul. 6 vols. Paris: 1857–82.
Cartulaire de l'abbaye de Berdoues-près-Mirande, ed. l'abbé Cazaurin. The Hague: Nijhoff, 1905.
Cartulaire de l'abbaye de Bonnecombe, ed. P.-A. Verlaguet. Rodez: Carrère, 1918–25.
Cartulaire de l'abbaye de Bonneval-en-Rouergue, ed. P.-A. Verlaguet, intro. J.-L. Rigal. Rodez: Carrère, 1938.
Cartulaire de l'abbaye de Cadouin, précédé de notes sur l'histoire économique du Périgord méridional à l'époque féodale, ed. J.-M. Maubourguet. Cahors: Couselant, 1926.
Cartulaire de l'abbaye de Gimont, ed. A. Clergeac. Paris: Champion, 1905.
Cartulaire de l'abbaye Notre-Dame de Bonnevaux au diocèse de Vienne, ordre de Cîteaux, ed. Ulysse Chevalier. Grenoble: F. Allier, 1889.
Cartulaire de l'abbaye Notre-Dame de Bonnevaux au diocèse de Vienne, ordre de Cîteaux, ed. anon. moine de Tamié. Notre-Dame de Tamié, Savoy, 1942.
Cartulaire de l'abbaye Notre-Dame de Léoncel, diocèse de Die, ordre de Cîteaux, ed. Ulysse Chevalier. Montélimar: Bourron, 1869.
Cartulaire de l'abbaye de Silvanès, ed. P.-A. Verlaguet. Rodez: Carrère, 1910.
Cartulaire de Trinquetaille, ed. P.-A. Amargier. Gap: Ophyrys, 1972.
Cartulaire et documents de l'abbaye de Nonenque, ed. C. Couderc and J.-L. Rigal. Rodez: Carrère, 1955.
Cartulaires des Templiers de Douzens, ed. Pierre Gérard and Elisabeth Magnou. Paris: Bibliothèque Nationale, 1965.
The Charters of St-Fursy of Péronne, ed. William Mendel Newman and Mary A. Rouse, preface by John F. Benton. Cambridge, Mass.: Medieval Academy, 1977.
Chartes et documents de l'abbaye de Notre-Dame d'Aiguebelle, Commission d'histoire de l'Ordre de Cîteaux II, ed. anon. 2 vols. Lyons: Audin, 1954, 1969.
Chartes de l'Ordre de Chalais: 1101–1401, ed. Jean-Charles Roman d'Amat. 3 vols. Paris: A. Picard et fils, 1923.
Chartes et documents concernant l'abbaye de Cîteaux 1098–1182, ed. J.-M. Marilier, Rome: Editions cisterciennes, 1961.
Chevalier, Jules. *L'abbaye de Notre-Dame de Valcroissant de l'ordre de Cîteaux au diocèse de Die*. Grenoble: Allier, 1897.
Documents concernants l'abbaye de Locdieu, ed. anon. Villefranche: Société des Amis, 1892.
Documents historiques sur le Tarn-et-Garonne, ed. François Moulenq. 4 vols. Montauban: Forestié, 1879–1894.
Documents sur l'ancien hôpital d'Aubrac, vol. I, ed. J.-L. Rigal and P.-A. Verlaguet. Rodez: Carrère, 1913–17. Vol. II, ed. J.-L. Rigal. Millau: Artières et Maury, 1934.

Gallia Christiana in provincias ecclesiasticas distributa, ed. Denys de Ste. Marthe. 16 vols. Paris: ex typographia regis, 1716–1865.
Gerald of Wales, *The Journey through Wales,* trans. Lewis Thorpe. Harmondsworth: Penguin, 1978.
Histoire générale de Languedoc, ed. Cl. Devic and J. Vaissete, 18 vols. Toulouse: Privat. 1872–1983, especially, "Inventaire des titres de Boulbonne," vol. 8, cols. 1883–1926, and "Inventaire des titres de Grandselve," vol. 8, cols. 1747–1803.
Idung de Prufening, *Cistercians and Cluniacs, the case for Cîteaux,* ed. and trans. Jeremiah F. O'Sullivan, Joseph Leaney and Grace Perrigo. Kalamazoo, Mich.: Cistercian Publications, 1977.
Layettes du Trésor des Chartes, ed. Alexandre Teulet, et al. 5 vols. Paris: Plon, 1863–1909.
Liber donationum Altaeripae: Cartulaire de l'abbaye cistercienne d'Hauterive (XIIe–XIIIe siècles), ed. Ernst Tremp, trans. Isabelle Bissegger-Garin. Lausanne: Société d'histoire de la Suisse romande, 1984.
Liber instrumentorum memorialium: Cartulaire des Guillems de Montpellier, ed. A. Germain. Montpellier, Jean Marel Ainé, 1884–1880.
Le premier cartulaire de l'abbaye cistercienne de Pontigny (xiie–xiiie siècles), ed. Martine Garrigues. Paris: Bibliothèque Nationale, 1981.
Recueil des actes de l'abbaye cistercienne de Bonnefont-en-Comminges, ed. Ch. Samaran and Ch. Higounet. Paris: Bibliothèque Nationale, 1970.
Recueil des pancartes de l'abbaye de la Ferté-sur-Grosne: 1113–1178, ed. Georges Duby. Aix-Marseilles: Publication de la faculté des lettres, 1953.
Statuta capitulorum generalium ordinis cisterciensis ab anno 1116 ad annum 1786, ed. J.-M. Canivez. 8 vols. Louvain: Bureaux, 1933.
The Tax Book of the Cistercian Order, ed. Arne Odd Johnson and Peter King. Oslo: Universitetsforlaget, 1979.
La vie de Saint Etienne d'Obazine, ed. Michel Aubrun. Clermont-Ferrand: Institut d'études du Massif Central, 1970.

SECONDARY SOURCES

Alart, B. "Monastères de l'ancien diocèse d'Elne, l'abbaye de Sainte-Marie de Jau ou de Clariana, ordre de Cîteaux." *Mémoires de la Société Agricole des sociétés des Pyrénées-Orientales* 11 (1872): 278–308.
Albe, le chanoine. *Les possessions de l'abbaye d'Obasine dans le diocèse de Cahors et les familles du Quercy. Titres et documents sur le Limousin et le Quercy.* Brives: Roche. 1911.
Alexis, dom. "La filiation de Clairvaux et l'influence de Saint Bernard au XIIe siècle." *Saint Bernard et son temps. Recueil des mémoires et communications présentés au congrès.* 2: 9–12. Dijon: L'Académie des sciences, arts et belles-lettres de Dijon, 1929.
Alibert, Louis. *Dictionnaire Occitan-Français d'après les parlers languedoçiens.* Toulouse: Institut d'études occitanes, 1966.
Angleton, Cicely d'A. "Two Cistercian Preaching Missions to Languedoc in the Twelfth Century: 1145, 1178." Unpublished Ph.D. dissertation, Catholic University of America, 1984.
D'Arbois de Jubainville, Henri. *Etudes sur l'état intérieur des abbayes cisterciennes, et principalement de Clairvaux, au XIIe et au XIIIe siècle.* Paris: August Durand Librairie, 1858.
Aubert, Marcel, with la Marquise de Maillé. *L'architecture cistercienne en France.* Paris: Editions d'art et d'histoire, 1943.
Baker, Derek. "Popular Piety in the Lodèvois in the Early Twelfth Century: The Case of Pons de Léras." *Religious Motivation: Biographical and Sociological Problems for the Church Historian. Studies in Church History* 15 (1978): 39–47.
Barley, M. W. "Cistercian Land Clearance in Nottinghamshire: Three Deserted Villages and their Moated Successor." *Nottingham Medieval Studies* 1 (1957): 75–89.
Barrau, H. de. "Etude historique sur l'ancienne abbaye de Bonnecombe." *Mémoires de la société des lettres, sciences, et arts de l'Aveyron* 2 (1839–40): 193–264.
Barrière, Bernadette. *L'abbaye cistercienne d'Obazine en Bas-Limousin.* Tulle: Imprimerie Orfeuil, 1977.
———. "Les granges de l'abbaye cistercienne d'Obazine au XIIe et XIIIe siècles." *Bulletin de la société des lettres, sciences, et arts de la Corrèze* 70 (1966): 33–51.
———. "L'économie cistercienne du sud-ouest de la France." *L'économie cistercienne. Géographie—*

Mutations du Moyen Age aux Temps modernes. Publications de la Commission d'histoire de Flaran, 3 (Auch: Comité départemental du Tourisme du Gers, 1983), pp. 75–100.
Barruol, Jean. *Sénanque et le pays du Lubéron au Ventoux*. Lyon: Lescuyer, 1975.
Batanay, J. "Les convers chez quelques moralistes des XII⁰ et XIII⁰ siècles." *Cîteaux: comm. cist.* 20 (1969): 241–59.
Bautier, Anne-Marie. "Les plus anciennes mentions de moulins hydrauliques industriels et de moulins à vent." *Bulletin Philologique et Historique (jusqu'à 1610) du Comité des Travaux Historiques et Scientifiques* 1 (1960): 567–625.
Bautier, Robert-Henri. *The Economic Development of Medieval Europe*, trans. Heather Karolyi. London: Harcourt, Brace and Jovanovich, 1971.
Beaunier, dom and J.-M. Besse. *Abbayes et prieurés de l'ancienne France*. 4 vols. Paris: Librairie Poussielgue, 1910.
Becquet, Jean. "L'érémitisme clérical et laïque dans l'Ouest de la France." In *Eremitismo in Occidente nei secoli XI e XII*. Atti della seconda Settimana internazionale di studio: Mendola, 30 agosto—6 settembre 1962, pp. 182–211. Milan: (Università Cattolica del Sacro Cuore, 1965).
Benouville, P., and P. Lauzun. "L'abbaye de Flaran." *Revue de Gascogne* 28 (1887): 575–81; 29 (1888): 291–302, 504–18; 30 (1889): 115–21, 221–33, 401–24.
Berenguier, Raoul. *Thoronet Abbey*, trans. John Seabourne. Paris: Caisse Nationale des Monuments Historiques, 1965.
Beresford, Maurice. *New Towns of the Middle Ages: Town Plantation in England, Wales, and Gascony*. New York: Praeger, 1967.
Berlière, Ursmer. "Les Origines de Cîteaux." *Revue d'histoire ecclésiastique* 1 (1900): 448–71; 2 (1901): 253–89.
Berman, Constance. "Administrative Records for the Cistercian Grange: the Evidence of the Cartularies of Bonnecombe." *Cîteaux: comm. cist.* 30 (1979): 201–20.
———. "Cistercian Development and the Order's Acquisition of Churches and Tithes in Southern France." *Revue bénédictine* 91 (1981): 193–203.
———. "Early Cistercian Expansion in Provence." *Heaven on Earth: Studies in Medieval Cistercian History IX*, ed. E. Rozanne Elder, pp. 43–54. Kalamazoo, Mich.: Cistercian Publications, 1983.
———. "Examples from the Diocese of Chartres: Economic Activities of Medieval Cistercian Houses for Women," forthcoming.
———. "Fortified Monastic Granges in the Rouergue." In *The Medieval Castle: Romance and Reality. Medieval Studies at Minnesota I*, ed. Kathryn Reyerson and Faye Powe, pp. 124–46. Dubuque, Ia.: Kendall and Hunt, 1984.
———. "The Foundation and Early History of the Monastery of Silvanès: the Economic Reality." In *Cistercian Ideals and Reality. Studies in Medieval Cistercian History III*, ed. J. R. Sommerfeldt, pp. 280–318. Kalamazoo, Mich.: Cistercian Publications, 1978.
———. "The Growth of the Cistercian Order in Southern France." *Analecta Cisterciensia*, forthcoming.
———. "Land Acquisition and the Use of the Mortgage Contract by the Cistercians of Berdoues." *Speculum* 57 (1982): 250–66.
———. "Men's Houses, Women's Houses: the Relationship between the Sexes in Twelfth-century Southern France," forthcoming.
———. "Monastic Hospices in Southern France: the Cistercian Urban Presence," forthcoming.
Bernard, Félix. *L'abbaye de Tamié: Ses granges: 1132-1793*. Grenoble: Allier, 1967.
Bernard de Clairvaux, préface de Thomas Merton. Commission d'histoire de Cîteaux, no. 3. Paris: Editions Alsatia, 1953.
Berthet, B. "Abbayes et exploitations. L'exemple de Saint-Claude et des forêts jurassiennes." *Annales: E.S.C.* 5 (1950): 68–74.
Bertrand, Georges. "Pour une histoire écologique de la France rurale." In *Histoire de la France rurale*, ed. Georges Duby and Armand Wallon, vol. 1, pp. 83–94. Paris: Seuil, 1975.
Bienvenu, J.-M. *L'étonnant fondateur de Fontévraud: Robert d'Arbrissel*. Paris: Nouvelles Editions Latines, 1981.
Bisson, Thomas N. *The Conservation of Coinage. Monetary Exploitation and its Restraint in France, Catalonia, and Arago (c. A.D. 1000-c. 1225)*. Oxford: Clarendon Press, 1979.
———. "The Organized Peace in Southern France and Catalonia, ca.1140-ca.1233." *American Historical Review* 82 (1977): 290–311.
Bligny, Bernard. *L'église et les ordres religieux dans la royaume de Bourgogne aux XI⁰ et XII⁰ siècles*. Grenoble: Allier, 1960.

Bloch, Marc. "The Advent and Triumph of the Watermill." In *Land and Work in Medieval Europe: Selected Papers*, trans. J. E. Anderson, pp. 136–68. New York: Harper and Row, 1969.
——. *Les caractères originaux de l'histoire rurale française*. Cambridge, Mass.: Harvard, 1931, and 2nd ed. with suppl. vol. II, by R. Dauvergne, drawn from Bloch's work: Paris: A. Colin, 1968. Produced in English as *French Rural History*, trans. Janet Sondheimer. Berkeley: University of California Press, 1970.
——. *Seigneurie française et manoir anglais*. Paris: A. Colin, 1967.
Bois, Michele. "La Basilique Sainte Anne." *Cîteaux dans la Drôme: Revue Drômoise* 83 (1980): 75–81.
Bonnassie, Pierre. "Les conventions féodales dans la Catalogne du XIe siècle." In *Les structures sociales de l'Aquitaine, du Languedoc, et de l'Espagne au premier âge féodale*. Colloques Internationaux du Centre National de la Recherche Scientifique: Toulouse, 28–31 mars, 1968. Paris: C.N.R.S., 1969.
Bouchard, Constance B. "Noble Piety and Reformed Monasticism: The Dukes of Burgundy in the Twelfth Century." In *Noble Piety and Reformed Monasticism. Studies in Medieval Cistercian History. VII*, ed. E. Rozanne Elder, pp. 1–9. Kalamazoo, Mich.: Cistercian Publications, 1981.
Bourgeat, Charles. *L'abbaye de Bouillas. Histoire d'une ancienne abbaye cistercienne du diocèse d'Auch*. Auch: F. Cocharaux, 1954. Also published as "L'abbaye de Bouillas." *Bulletin de la société archéologique, historique et scientifique de Gers* (Auch) 54 (1953): 273–82; 55 (1955): 60–90, 388–405; 56 (1956): 83–109, 180–203.
Bourgeois, Giselle. "Les granges et l'économie de l'abbaye de Nonenque au Moyen Age." *Cîteaux: comm. cist.* 24 (1973): 139–60.
Bousquet, l'abbé. "Les anciennes abbayes de l'ordre de Cîteaux dans le Rouergue." *Mémoires de la société des lettres, sciences, et arts de l'Aveyron* 9 (1867): 9–29.
Bousquet, Jacques. "Le traité d'alliance entre Hughes, comte de Rodez et les consuls de Millau: 6 juin 1223." *Annales du Midi* 72 (1960): 24–42.
——. "Les origines de la transhumance en Rouergue." *L'Aubrac: Etude ethnologique, linguistique, agronomique, et économique d'un établissement humain*, vol. 2, pp. 217–55. Paris: C.N.R.S., 1971.
Boussard, J. "La vie en Anjou aux XIe et XIIe siècles." *Le Moyen Age* 56 (1950): 50–53.
Boutruche, Robert. *Une société provinciale en lutte contre le régime féodal. L'alleu en Bordelais et en Bazadais du XIe au XVIIIe siècle*. Rodez: P. Carrère, 1943.
Boyd, Catherine E. *A Cistercian Nunnery in Medieval Italy: The Story of Rifreddo in Saluzzo, 1220–1300*. Cambridge, Mass.: Harvard University Press, 1943.
——. *Tithes and Parishes in Medieval Italy: The Historical Roots of a Modern Problem*. Ithaca, N.Y.: Cornell University Press for the American Historical Association, 1952.
Braudel, Fernand. *The Mediterranean and the Mediterranean World in the Age of Philip II*, trans. Sian Reynolds. 2 vols. New York: Harper, 1972.
Brelot, J. "La fondation et le développement des abbayes cisterciennes dans le comté de Bourgogne au XIIe siècle." *Mémoires de la société pour l'historie du droit et des institutions des anciens pays bourguignons, comtois, et romands* 15 (1953): 133–52.
Brown, E. A. R. "The Cistercians in the Latin Empire of Constantinople and Greece, 1204." *Traditio* 14 (1958): 63–120.
Brutails, J.-A. "De l'ancienne organisation de la propriété territoriale dans le Midi de la France." *Revue des Pyrénées (Toulouse)* 3 (1891): 154–71.
Bruzelius, Caroline A. "Cistercian High Gothic: the Abbey Church of Longpont and the Architecture of the Cistercians in the Early Thirteenth Century," *Analecta Cisterciensia* 35 (1979): 3–204.
——. "The Twelfth-Century Church at Ourscamp." *Speculum* 56 (1981): 28–40.
Cabanis, Alfred. "L'abbaye de Valmagne." *Mémoires de la société des lettres, sciences, et arts de l'Aveyron* 5 (1844–45): 424–27.
Canivez, J.-M. "Beaulieu-en-Rouergue." *Dictionnaire d'histoire et de géographie ecclésiastique* 7 (1934): 164.
——. "Belleperche." *Dictionnaire d'histoire et de géographie ecclésiastique* 8 (1936): 869–71.
——. "Bonnevaux." *Dictionnaire d'histoire et de géographie ecclésiastique* 9 (1937): 1074–76.
——. "Calers." *Dictionnaire d'histoire et de géographie ecclésiastique* 11 (1949): 387–88.
——. "Candeil." *Dictionnaire d'histoire et de géographie ecclésiastique* 11 (1949): 719.
——. "Cîteaux." *Dictionnaire d'histoire et de géographie ecclésiastique* 12 (1953): 852–997.
Carville, G. "Cistercian Mills in Medieval Ireland." *Cîteaux. comm. cist.* 24 (1973): 310–18.

———. "The Urban Property of the Cistercians in Medieval Ireland." *Studia monastica* 14 (1972): 35–47.
Castaing-Sicard, Mireille. *Les contrats dans le très ancien droit toulousain (x^e–xiiie siècle)*. Toulouse: Espic, 1957.
———. "Donations toulousaines du Xe au XIIIe siècles." *Annales du Midi* 70 (1958): 27–64.
———. *Monnaies féodales et circulation monétaire en Languedoc (x^e–xiiie siècles)*. Toulouse: Association Marc Bloch, 1961.
Castelnau, L. de. "L'abbaye de Bonneval." *Revue du Midi* (1887): 65–69.
Cathala, C. "Boulbonne." *Dictionnaire d'histoire et de géographie ecclésiastique* 10 (1938): 67–70.
Cauvet, E. *Etude historique sur Fonfroide*. Montpellier: Séquin, 1975.
Cavailles, Henri. *La transhumance pyrénéenne et la circulation des troupeaux dans les plaines de Gascogne*. Paris: Université de Paris, 1931.
———. *La vie pastorale et agricole dans les Pyrénées des Gaves, de l'Adour, et des Nestes*. Paris: Armand Colin, 1931.
Chaigne, Léon. "L'implantation cistercienne en Poitou." *Bulletin de la société d'émulation de la Vendée* 120 (1971): 11–19.
Champier, L. "Cîteaux: Ultime étape dans l'aménagement agraire de l'occident." *Mélanges Saint Bernard. XXIVe congrès de l'association bourguignonne des sociétés savantes*, pp. 254–61. Dijon: Association des amis de Saint Bernard, 1954.
Chauvin, Benoît. "Cisterciensia." *Francia. Forschungen zur Westeuropäischen Geschichte* 10 (1982): 514–20.
———. "De la 'Villa Carolingienne' à la grange cistercienne: le cas de la terre de Glénon (Arbois, Jura) du milieu de X^e à la fin du XIIe siècle." *Villa—Curtis—Grangia. Landwirtschaft zwischen Loire und Rhein von der Römerzeit zum Hochmittelalter. Economie Rurale entre Loire et Rhin de l'Epoque Gallo-Romaine au XIIe- XIIIe siècle*. Beihefte der Francia, 11, ed. Walter Janssen and Dietrich Lohrmann, preface by Charles Higounet, pp. 164–84. Munich: Artemis Verlag, 1983.
———. "Réalités et évolution de l'économie cistercienne dans les duché et comté de Bourgogne au Moyen Age. Essai de synthèse." *L'économie cistercienne. Géographie—Mutations du Moyen Age aux Temps modernes*. Publications de la Commission d'histoire de Flaran, 3 (Auch: Comité départemental du Tourisme du Gers, 1983), pp. 13–52.
Chédeville, André. *Chartres et ses campagnes (XIe–XIIIe siècles)*. Paris: Klincksiek, 1973.
———. "Etude de la mise en valeur et du peuplement du Maine au XIe siècle." *Annales de Bretagne* 67 (1960): 209–25.
Cheney, C. R. "A Letter of Pope Innocent III and the Lateran Decree on Cistercian Tithe-paying." *Cîteaux: comm. cist.* 13 (1962): 146–51.
Chevanne, J. de. "Le site primitif de l'abbaye de Maizières, la paroisse disparue de La Bretenière, et le déplacement général des abbayes cisterciennes en Bourgogne." *Mémoires de la société pour l'histoire du droit et des institutions des anciens pays bourguignons, comtois, et romands* 15 (1953): 59–64.
Chevrier, G. "Remarques sur la distinction de l'acte à titre onéreux et de l'acte à titre gratuit d'après les chartes du Rouergue au XIIe siècle." *Annales de la faculté de droit de l'Université d'Aix-Marseilles* 43 (1950): 67–79.
Cheyette, Frederic. "European Villages." *Journal of Economic History* 37 (1977): 198–209.
Clément, Paul. "Les monastères cisterciens de Suisse romande." *Mémoires de la société pour l'histoire du droit et des institutions des anciens pays bourguignons, comtois, et romands* 15 (1953): 183–99.
Clergeac, A. "Les abbayes de Gascogne du XIIe siècle au grand schisme d'occident." *Revue de Gascogne* 47 (1906): 316–29, 530–44.
Cocheril, Maur. "Espagne cistercienne." *Dictionnaire d'histoire et de géographie ecclésiastique* 15 (1963): 944–69.
———. "L'implantation des abbayes cisterciennes dans la peninsule ibérique." *Annuario des Estudios Medievales* 1 (1964): 217–87.
———. "Les monastères cisterciennes du nord du Portugal." *Collectanea ordinis cisterciensium reformatorum* 19 (1957): 163–82, 355–70.
Comba, Rinaldo. "Aspects économiques de la vie des abbayes cistercienne de l'Italie du Nord-Ouest (XIIe–XIVe siècle)." *L'économie cistercienne. Géographie—Mutations du Moyen Age aux Temps modernes*. Publications de la Commission d'histoire de Flaran, 3 (Auch: Comité départemental du Tourisme du Gers, 1983), pp. 119–134.

Constable, Giles. "Cluniac Tithes and the Controversy between Gigny and Le Miroir." *Revue bénédictine* 70 (1960): 591-624.
———. "*Famuli* and *Conversi* at Cluny." *Revue bénédictine* 83 (1973): 326-50.
———. *Medieval Monasticism. A Select Bibliography*. Toronto: University of Toronto Press, 1976.
———. *Monastic Tithes from their Origins to the Twelfth Century*. Cambridge: Cambridge University Press, 1964.
———. "The Study of Monastic History Today." In *Essays on the Reconstruction of Medieval History*, ed. Vaclav Mudroch and G. S. Couse, pp. 21-51. Montreal and London, Ontario: McGill and Queen's University Press, 1974.
Coste, Pierre. "La vie pastorale en Provence au milieu du XIVe siècle." *Etudes rurales* 46 (1972): 61-75.
Cottineau, L.-H. *Répertoire topo-bibliographique des abbayes et prieurés*. Macon: Protat frères, 1935-38.
Croix-Bouton, Jean de la. "Le repeuplement des baillages de Reauville et de Grignan au XVe siècle." *Cîteaux dans la Drôme: Revue Drômoise* 83 (1980): 55-59.
Crozet, René. "L'épiscopat de France et l'ordre de Cîteaux au XIIe siècle." *Cahiers de civilisation médiévale* 18 (1975): 263-68.
Cuerne, Louis de. "L'abbaye cistercienne de Locdieu." *Connaissance du Monde* 84 (1965): 13-27.
Cursente, B. *Les castelnaux de la Gascogne médiévale*. Bordeaux: Fédération historique du Sud-Ouest, 1980.
David-Roy, Marguerite. "Les granges monastiques en France aux XIIe et XIIIe siècles." *Archeologia* 58 (1973): 53-62.
Davis, Natalie Zemon. *The Return of Martin Guerre*. Cambridge, Mass.: Harvard University Press, 1983.
De Keyzer, J. Walter. "La limite entre les polders de Lille et de Berendrecht au XIIIe siècle." In *Hommage au Professeur Bonenfant. 1899-1965. Etudes d'histoire médiévale dédiées à sa mémoire par les anciens élèves de son seminaire à l'université libre de Bruxelles*, pp. 221-29. Brussels: G. Despy, 1965.
Delahaye, Gilbert-Robert. "La prise de possession de la forêt d'Echou (Seine-et-Marne) par les religieux cisterciens de Preuilly aux XIIe et XIIIe siècles." *Paris et Ile-de-France. Mémoires publiés par la fédération des sociétés historiques et archéologiques de Paris et de l'Ile-de-France* 28 (1978): 85-96.
Delessard, L. "Les débuts de l'abbaye de Morimond." *Mémoires de la société pour l'histoire du droit et des institutions des anciens pays bourguignons, comtois, et romands* 15 (1953): 65-68.
Desmarchelier, M. "Quelques notes sur l'abbaye cistercienne de Trizay (Vendée)." *Cîteaux: comm. cist.* 25 (1974): 33-60.
Dijk, C. van. "L'instruction et la culture des frères convers dans les premiers siècles de l'ordre de Cîteaux." *Collectanea ordinis cisterciensium reformatorum* 24 (1962): 243-58.
Dimier, M.-A. *L'art cistercien*, trans. Paul Veyriras. 2nd ed. Paris: Zodiaque, 1974.
———. "Cîteaux et les emplacements malsains." *Cîteaux in de Nederlanden* 6 (1955): 89-97.
———. "Eaunes." *Dictionnaire d'histoire et de géographie ecclésiastique* 14 (1963): 1263-1266.
———. "L'Escaledieu." *Dictionnaire d'histoire et de géographie ecclésiastique* 15 (1963): 1844.
———. "Feuillans." *Dictionnaire d'histoire et de géographie ecclésiastique* 16 (1967): 1334.
———. "Granges, celliers, et bâtiments d'exploitation cisterciens." *Archeologia* 74 (1974): 46-57.
———. "Quelques légendes de fondations chez les cisterciens." *Studia monastica* 12 (1970): 97-105.
Donkin, R. A. *A Check-list of Printed Works Relating to the Cistercian Order. Documentation cistercienne. II*. Rochefort, Belgium: Abbaye N. D. de S. Rémy, 1969.
———. "The Cistercian Grange in England in the Twelfth and Thirteenth Centuries, with Special Reference to Yorkshire." *Studia monastica* 6 (1964): 95-144.
———. "The Cistercian Order in Medieval England: Some Conclusions." *Institute of British Geographers: Transactions* 33 (1963): 296-310.
———. "The Cistercian Settlement and the English Royal Forests." *Cîteaux: comm. cist.* 11 (1960): 39-55, 117-32.
———. *The Cistercians: Studies in the Geography of Mediaeval England and Wales*. Toronto: Pontifical Institute of Mediaeval Studies, 1978.
———. "The Disposal of Cistercian Wool in England and Wales during the Twelfth and Thirteenth Centuries." *Cîteaux: comm. cist.* 8 (1957): 1099-1131; 9 (1958): 181-202.

———. "The English Cistercians and Assarting, c.1128–c.1350." *Analecta sacri ordinis cisterciensis* 20 (1964): 49–75.

———. "The Growth and Distribution of the Cistercian Order in Medieval Europe." *Studia monastica* 9 (1967): 275–86.

———. "The Markets and Fairs of Medieval Cistercian Monasteries in England and Wales." *Cistercienser-chronik* 69 (1962): 1–14.

———. "The Marshland Holdings of the English Cistercians before c.1350." *Cîteaux: comm. cist.* 9 (1958): 262–75.

———. "Settlement and Depopulation on Cistercian Estates during the Twelfth and Thirteenth Centuries especially in Yorkshire." *Bulletin of the Institute of Historical Research* 33 (1960): 141–65.

———. "The Site Changes of Mediaeval Cistercian Monasteries." *Geography* 44 (1959): 251–58.

———. "Some Aspects of Cistercian Sheep Farming in England and Wales." *Cîteaux: comm. cist.* 13 (1962): 296–310.

———. "The Urban Property of the Cistercians in Mediaeval England." *Analecta cisterciensia* 15 (1959): 104–31.

Donnelly, James S. "Changes in the Grange Economy of English and Welsh Cistercian Abbeys, 1300–1540." *Traditio* 10 (1954): 399–458.

———. *The Decline of the Medieval Cistercian Lay-Brotherhood.* Fordham University Studies in History: 3. New York: Fordham University, 1949.

Donnet, N. "La fondation de l'abbaye d'Argensolles." *Cîteaux: comm. cist.* 10 (1959): 212–18.

Douais, M. "Le cartulaire de Nizors." *Revue de Gascogne* 28 (1887): 565–67.

Dubled, Henri. "Aspects de l'économie cistercienne en Alsace." *Revue d'histoire ecclésiastique* 54 (1959): 765–82.

Dubois, Jacques. "L'institution des convers au XIIe siècle. Forme de vie monastique propre aux laïcs." *I laici nella "societas Christiana" dei secoli XI e XII.* Atti della terza Settimana internazionale di studio, Mendola, 21–27 agosto, 1965, pp. 183–261. Milan: Università Cattolica del Sacro Cuore, 1968.

———. "Les moines dans la société du Moyen Age: 950–1350." *Revue d'histoire de l'église de France* 60 (1974): 5–37.

Dubord, R. "Les abbayes cisterciennes filles de Gimont." *Revue de Gascogne* 27 (1876): 221–22.

———. "Essai historique sur l'abbaye de Gimont." *Revue de Gascogne* 11 (1871): 427–31; 12 (1872): 93–101, 193–204, 289–302; 13 (1873): 49–67, 227–43; 14 (1874): 25–38, 69–84, 448–56, 496–505; 15 (1875): 69–80, 156–74.

Dubourg, A. *Ordre de Malte. Histoire du Grand Prieuré de Toulouse.* Paris: Société bibliographique, 1883.

Duby, Georges. "Le budget de l'abbaye de Cluny." In *Hommes et structures du Moyen Age. Recueil d'articles,* pp. 61–82. Paris: Mouton, 1973.

———. *Bernard de Clairvaux: L'art cistercien.* Paris: Arts et Métiers, 1976.

———. *The Early Growth of the European Economy,* trans. Howard B. Clarke. Ithaca, N.Y.: Cornell University Press, 1974.

———. "Un inventaire des profits de la seigneurie clunisiene à la mort de Pierre le Vénérable." In *Hommes et structures du Moyen Age. Recueil d'articles,* pp. 87–101. Paris: Mouton, 1973.

———. *Rural Economy and Country Life in the Medieval West,* trans. Cynthia Postan. Columbia, S.C.: University of South Carolina Press, 1968.

———. *La société aux XIe et XIIe siècles dans la région mâconnaise.* Paris: S.E.V.P.E.N., 1971.

——— and Armand Wallon, eds. *Histoire de la France Rurale.* 5 vols. Paris: Seuil, 1975.

DuCange, Charles de Fresne. *Glossarium mediae et infimae latinitatis,* ed. Léopold Favre. 10 vols. Paris: Librairie des sciences et des arts, 1937–38.

Dupont, André. "L'exploitation du sel sur les étangs de Languedoc, IXe–XIIIe siècle." *Annales du Midi* 70 (1958): 7–25.

Durand, Robert. "L'économie cistercienne au Portugal." *L'économie cistercienne. Géographie—Mutations du Moyen Age aux Temps modernes. Publications de la Commission d'histoire de Flaran,* 3 (Auch: Comité départemental du Tourisme du Gers, 1983), pp. 101–118.

Durliat, Marcel. "L'abbaye de Villelongue." In *Carcassonne et sa Region. Actes des XLIe et XXIVe congrès d'études régionales tenus par la Fédération historique du Languedoc méditerranéen et du Roussillon et par la Fédération des Sociétés académiques et savantes de Languedoc-Pyrénées-Gascogne,* pp. 65–67. Carcassonne: C.N.R.S., 1970.

Duval-Arnould, Louis. "Les ressources de la forêt royale de Retz et leur place dans l'économie de l'abbaye de Longpont." *L'économie cistercienne. Géographie—Mutations du Moyen Age aux Temps modernes. Publications de la Commission d'histoire de Flaran*, 3 (Auch: Comité départemental du Tourisme du Gers, 1983), pp. 189–96.

Duvivier, C. "Hospités. Défrichements en Europe et spécialement dans nos contrées aux XIe, XIIe, et XIIIe siècles." *Revue d'histoire et d'archéologie* (Brussels) 1 (1859): 74–90, 131–75.

Ebersolt, J. G. "La seigneurie et la société dans les chartes de l'abbaye cistercienne d'Acey au XIIe siècle." *Mémoires de la société pour l'histoire du droit et des institutions des anciens pays bourguignons, comtois, et romands* 15 (1953): 153–72.

Eckenrode, T. R. "The English Cistercians and their Sheep During the Middle Ages." *Cîteaux: comm. cist.* 24 (1973): 250–66.

Emery, Richard W. *Heresy and Inquisition in Narbonne*. New York: Columbia University Press, 1941.

Enjalbert, Henri. *Rouergue Quercy*. Paris: Arthaud, 1971.

Evans, Austin P. "The Albigensian Crusade." In *A History of the Crusades*, ed. Kenneth M. Setton. vol. 2. *The Later Crusades, 1189–1311*, ed. Robert Lee Wolff, and Harry W. Hazard. Philadelphia: University of Pennsylvania Press, 1962.

Fel, André. "Réflexions sur les paysages agraires des hautes terres du Massif Central français." *Géographie et histoire agraires. Actes du colloque international organisé par la Faculté des Lettres de l'Université de Nancy: 2–7 Septembre, 1957. Annales de l'Est, Mémoire* no. 21 (1959): 155–67.

Fénelon, Paul. "Structure des finages périgourdins." *Géographie et histoire agraires. Actes du colloque international organisé par la Faculté des Lettres de l'Université de Nancy: 2–7 Septembre, 1957. Annales de l'Est, Mémoire* no. 21 (1959): 168–92.

Ferras, Vincent. *Un cistercien occitan au XIIe siècle: Pons de Léras: Lodêvois et fondateur du monastère de Sylvanès*. Dourgne: Abbaye d'En Calcat, 1971.

———. *Documents bibliographiques concernant le rayonnement médiéval de l'Ordre de Cîteaux en pays d'Aude*. Dourgne: Abbaye d'En Calcat, 1971.

Feuchère, M. "Le défrichement des forêts en Artois du IXe au XIIIe siècle." *Bulletin trimestriel de la société académique des antiquaires de la Morinie* 18 (1952): 33–45.

Fons, Victor. "Les monastères cisterciens de l'ancienne province ecclésiastique de Toulouse." *Revue de Toulouse et du Midi de la France* 25 (1867): 112–34.

Foreville, Raymonde. "Arnaud Amalric, archévêque de Narbonne: 1196–1225." *Narbonne: Archéologie et Histoire.* XLVe Congrès organisé par la fédération historique du Languedoc Mediterranéen et du Roussillon, pp. 129–46. Montpellier: Fédération, 1973.

Fossier, Robert. "L'essor économique de Clairvaux." *Bernard de Clairvaux*, préface de Thomas Merton. Commission d'histoire de Cîteaux, no. 3, pp. 95–114. Paris: Editions Alsatia, 1953.

———. "La place des cisterciens dans l'économie picarde des XIIe et XIIIe siècles." *Mélanges historiques réunis à l'occasion du neuvième centenaire de l'abbaye d'Orval. Colloque, 1970*, pp. 273–81. Liège: Soledi, 1970.

———. *La terre et les hommes en Picardie*. 2 vols. Louvain: Nauwelaertz, 1968.

———. "L'économie cistercienne dans les plaines du nord-ouest de l'Europe." *L'économie cistercienne. Géographie—Mutations du Moyen Age aux Temps modernes. Publications de la Commission d'histoire de Flaran*, 3 (Auch: Comité départemental du Tourisme du Gers, 1983), pp. 53–74.

Fournier, Gabriel, "La création de la grange de Grégovie par les Prémontrés de Saint André et sa transformation en seigneurie (XIIe–XVIe siècles)." *Le Moyen Age* 56 (1950): 307–55.

———. *Le peuplement rural en Basse-Auvergne durant le haut Moyen Age*. Paris: Presses Universitaires de France, 1962.

Fourquin, Guy. "Le temps de la croissance." *Histoire de la France Rurale*, ed. Georges Duby and Armand Wallon, vol. 1, pp. 377–552. Paris: Seuil, 1975.

———. *Histoire économique de l'Occident médiévale*, 3rd ed. Paris: A. Colin, 1979.

———. *Le paysan de l'occident au moyen âge*. Paris: F. Nathan, 1972.

Framond, M. de. "Historique de l'abbaye de la Vernaison." *Cîteaux dans la Drôme: Revue Drômoise* 83 (1980): 151–54.

Froehly, W. "Les débuts du monastère de Theuley." *Mémoires de la société pour l'histoire du droit et des institutions des anciens pays bourguignons, comtois, et romands* 15 (1953): 173–76.

Gallagher, Philip F. "Conditions of Land Tenure and their Religious Implications at Twelfth-Century Mortemer." In *Studies in Medieval Cistercian History II*, ed. John R. Sommerfeldt, pp. 106–22. Kalamazoo, Mich.: Cistercian Publications, 1976.

Garrigues, D. "L'abbaye Notre-Dame d'Eaunes en Comminges." *Revue de Comminges* (St. Gaudens) 26 (1912): 133–50, 255–70; 28 (1915): 1–14.
Gatheron, M.-J. "Sur la continuité du role agraire des cisterciens." *Saint Bernard et son Temps. Recueil des mémoires et communications présentés au congrès*, 2: 89–94. Dijon: L'académie des sciences, arts et belles-lettres de Dijon, 1929.
Gaudemet, J. "Le diocèse de Langres au temps de Saint Bernard." *Mémoires de la société pour l'histoire du droit et des institutions des anciens pays bourguignons, comtois, et romands* 15 (1953): 69–81.
Gayne, P. "L'abbaye du Grandselve." *Bulletin de la société archéologique de Tarn et Garonne* 77 (1949): 104–27.
Geary, Patrick. *Furta Sacra: Theft of Relics in the Central Middle Ages*. Princeton: Princeton University Press, 1978.
Génestal, Robert. *Le rôle des monastères comme établissements de crédit: étudié en Normandie du XIe à la fin du XIIIe siècle*. Paris: Arthur Rousseau, 1901.
Gerards, A. "Die ersten Cistercienser und die Handarbeit." *Cistercienserchronik* 59 (1952): 1–6.
———. "Wirtschaftliche Hintergründe zur Zeit der Gründung des Cistercienser Ordens." *Cistercienserchronik* 58 (1951): 65–79.
Gimpel, Jean. *The Medieval Machine. The Industrial Revolution of the Middle Ages*. New York: Penguin, 1976.
Gorsse, Pierre de. *L'abbaye cistercienne Sainte-Marie de Valmagne au diocèse d'Agde en Languedoc*. Toulouse: Lion, 1933.
Gozzo, Francesco. *Vita Economica delle Abbazie Piemontesi (Secoli X–XIV)*. Rome: Analecta Gregoriana, 1940.
Grand, Roger. *Le contrat de complant dupuis les origines jusqu'à nos jours*. Paris: L. Tenin, 1917.
———. "Note d'économie agraire médiévale: 'Mansus Vestitus' et 'Mansus Absus.' " In *Etudes d'histoire du droit privé offertes à Pierre Petot*, pp. 251–55. Paris: Librairie générale de droit et de jurisprudence, 1959.
Graves, Coburn V. "The Economic Activities of the Cistercians in Medieval England. (1128–1307)." *Analecta sacri ordinis cisterciensis* 13 (1957): 3–60.
———. "Medieval Cistercian Granges." In *Studies in Medieval Culture* II, ed. J. R. Sommerfeldt, pp. 63–70. Kalamazoo, Mich.: Western Michigan University, 1966.
Grèzes-Rueff, François. "L'abbaye de Fontfroide et son domaine foncier au XIIe–XIIIe siècles." *Annales du Midi* 89 (1977): 253–80.
Grillon, L. "Deux granges corréziennes de l'abbaye cistercienne de Dalon." *Bulletin de la société des lettres, sciences et arts de la Corrèze* 70 (1966): 21–32.
Guillemain, Bernard. "Les moines sur les sièges épiscopaux du sud-ouest de la France aux XIe et XIIe siècles." *Etudes de civilisation médiévale IXe–XIIe siècles. Mélanges offerts à Edmund-René Labande*, pp. 377–84. Poitiers: C.E.S.C.M., 1975.
Guillot, Michel. "Ermites et ermitages des forêts d'Ile-de-France à l'âge classique." *Paris et Ile-de-France. Mémoires publiés par la fédération des sociétés historiques et archéologiques de Paris et de l'Ile-de-France* 28 (1978): 273–97.
Hallinger, K. "Woher kommen die laienbruder?" *Analecta cisterciensia* 12 (1956): 1–104.
Hays, Rhys W. "Further Notes on the Welsh Cistercians." *Cîteaux: comm. cist.* 22 (1971): 99–103.
Herlihy, David. "The Agrarian Revolution in Southern France and Italy, 801–1150." *Speculum* 33 (1958): 23–41.
———. "Church Property on the European Continent, 701–1200." *Speculum* 36 (1961): 81–105.
———. "Treasure Hoards in the Italian Economy, 960–1139." *Economic History Review* 10 (1957): 1–14.
Higounet, Charles. "Les Alamans, seigneurs bastidors et péagers du XIIe siècle." *Annales du Midi* 65 (1953): 227–54. [*Paysages*, pp. 305–24.]
———. "Les artigues des vallées luchonnaises." *France méridionale et pays ibériques. Mélanges géographiques offerts en hommage au Doyen D. Faucher*, 2: 555–82. Toulouse: Privat, 1949. [*Paysages*, pp. 83–104.]
———. "Une carte agricole de l'Albigeois vers 1260." *Annales du Midi* 70 (1958): 65–71. [*Paysages*, pp. 187–92.]
———. "Une carte des relations monastiques transpyrénéennes au Moyen Age." *Revue de Comminges* 64 (1951): 129–38.

———. "Cisterciens et bastides." *Le Moyen Age* 56 (1950): 69–84. [*Paysages*, pp. 265–74.]

———. "Contribution à l'étude de la toponymie du défrichement, les 'artigues' du Bordelais et du Bazadais." *IIIe Congrès international de toponymie et anthroponymie: Louvain: Actes et Mémoires* (1951), 3: 595–603. [*Paysages*, pp. 105–10.]

———. "Essai sur les granges cisterciennes." *L'économie cistercienne. Géographie—Mutations du Moyen Age aux Temps modernes. Publications de la Commission d'histoire de Flaran*, 3 (Auch: Comité départemental du Tourisme du Gers, 1983), 157–80.

———. "L'expansion de la vie rurale au XIIe et au XIIIe siècle." *Information historique* 16 (1953): 17–22.

———. "La grange de Champigny. Un terroir cistercien champenois à la fin du moyen âge," *Cîteaux: comm. cist.* 35 (1984): 83–91.

———. *La grange de Vaulerent: Structure et exploitation d'un terroir cistercien de la plaine de France: XIIe–XVe siècle*. Paris: S.E.V.P.E.N., 1965.

———. "Hagiotoponymie et histoire. Sainte Eulalie dans la toponymie de la France." *Actes et Mémoires du Cinquième Congrès international de Sciences Onomastiques*, 1: 105–13. Salamanca: Congrès, 1958. [*Paysages*, pp. 77–82.]

———. "Les hommes, la vigne et les églises romanes du Bordelais et du Bazadais." *Revue d'histoire de Bordeaux* 1 (1952): 107–11. [*Paysages*, pp. 111–18.]

———. "Un mémoire sur les péages de la Garonne du début du XIVe siècle." *Annales du Midi* 61 (1948–49): 320–24.

———. "Observations sur la seigneurie rurale et l'habitat en Rouergue, du IXe au XIVe siècle." *Annales du Midi* 62 (1950): 121–34. [*Paysages*, pp. 151–60.]

———. "L'occupation du sol du pays entre Tarn et Garonne au moyen âge." *Annales du Midi* 65 (1953): 301–30. [*Paysages*, pp. 129–50.]

———. *Paysages et villages neufs du Moyen Age: recueil d'articles de Charles Higounet*. Bordeaux: Fédération historique du Sud-Ouest, 1975.

———. "Le peuplement de Toulouse au XIIe siècle." *Annales du Midi* 55 (1943): 489–98.

———. "Le premier siècle de l'économie rurale cistercienne." *Istituzioni monastiche e istituzioni canonicali in occidente (1123–1215). Atti della settima Settimana internazionale di studio, Mendola, 28 agosto–3 settembre, 1977*, pp. 345–68. Milan: Vita e Pensiero, 1980.

———. "Les types d'exploitations cisterciennes et prémontrées et leur rôle dans la formation de l'habitat et des paysages ruraux." *Géographie et histoire agraires. Actes du colloque international organisé par la faculté des lettres de l'université de Nancy: 2–7 Septembre, 1957. Annales de l'Est, Mémoire no. 21*, (1959): 260–71. [*Paysages*, pp. 177–84.]

———. "Zur Siedlungsbewegungen Südwestfrankreichs vom 11. bis zum 14. Jahrhundert." *Die Deutsche Ostsiedlung des Mittelalters als problem der Europäische Geschichte. Reichenau: Vorträge, 1972*, ed. Walter Schlesinger, pp. 657–94. Sigmaringen: Thorbecke, 1975.

Hill, Bennett D. *English Cistercian Monasteries and their Patrons in the Twelfth Century*. Urbana: University of Illinois Press, 1968.

Hill, J. and H. Hill. *Raimond IV de Saint-Gilles, comte de Toulouse*. Toulouse: Privat, 1959.

Hoffmann, Eberhard. "Die Entwicklung der Wirtschaftsprinzipien im Cisterzienser-orden während des 12. und 13. Jahrhunderts." *Historisches Jahrbuch. Goerres-Gesellschaft* 31 (1910): 699–727.

———. *Das Konversenistitut des Cisterzienserordens in seinem Ursprung und seiner Organisation*. Freiburg: Verlag der Universitäts-Buchhandlung, 1905.

Hoffmann, Richard C. "The Medieval Origins of the Common Fields." In *European Peasants and their Markets*, ed. William N. Parker and Eric L. Jones, pp. 23–71. Princeton: Princeton University Press, 1975.

Horn, Walter and Ernest Born. "The Barn of the Cistercian Grange of Vaulerent (Seine-et Oise), France," In *Festschrift Ulrich Middeldorf*, ed. Antje Kosegarten and Peter Tigler, pp. 24–41. Berlin: Walter de Gruyter and Co., 1968.

Janauschek, Leopoldus. *Originum Cisterciensium* [Vienna, 1877] Englewood, N.J.: Gregg Press, 1964.

Janssen, Wilhelm. "Zisterziensische wirtschaftsführung am niederrhein: das Kloster Kamp und seine Grangien im 12.-13. Jahrhundert." *Villa—Curtis—Grangia. Landwirtschaft zwischen Loire und Rhein von der Römerzeit zum Hochmittelalter. Economie Rurale entre Loire et Rhin de l'Epoque Gallo-Romaine au XIIe–XIIIe siècle. Beihefte der Francia*, 11, ed. Walter Janssen and Dietrich Lohrmann, preface by Charles Higounet, pp. 205–21. Munich: Artemis Verlag, 1983.

Jobin, Isabelle. "Abbaye Notre-Dame de Valcroissant." *Cîteaux dans la Drôme: Revue Drômoise* 83 (1980): 134–48.
Johnsen, Arne Odd. *De norske cistercienserklostre: 1146–1264. Sett i europiesk sammenhang.* Oslo: Universitetsforlaget, 1977.
Johnson, Penelope D. *Prayer, Patronage, and Power. The Abbey of la Trinité, Vendôme, 1032–1187.* New York: New York University Press, 1981.
Jongler, M. "Monographie de l'abbaye de Grandselve." *Mémoires de la société archéologique du Midi* 7 (1853–60): 179–234.
Kempf, Jean-Pierre. *L'abbaye de Cherlieu XIIe XIIIe siècles: Economie et société.* Vesoul: Salsa, 1976.
Kloczowski, Jerzy. "Les Cisterciens en Pologne du XIIe au XIIIe siècle." *Cîteaux: comm. cist.* 21 (1970): 11–134.
Knowles, David. *Cistercians and Cluniacs: The Controversy between St. Bernard and Peter the Venerable.* London: Oxford University Press, 1955.
———. *The Monastic Order in England, 940–1216*, 2nd ed. Cambridge: Cambridge University Press, 1963.
———. "The Primitive Cistercian Documents." In *Great Historical Enterprises: Problems in Monastic History*, pp. 197–222. London: Thomas Nelson and Sons, 1963.
Lacger, Louis de. "Ardorel," *Dictionnaire d'histoire et de géographie ecclésiastique* 7 (1924): 1617–20.
Lackner, Bede. "Early Cîteaux and the Care of Souls." In *Noble Piety and Reformed Monasticism. Studies in Medieval Cistercian History VII*, ed. E. Rozanne Elder, pp. 52–67. Kalamazoo, Mich.: Cistercian Publications, 1981.
———. *The Eleventh-Century Background of Cîteaux.* Washington, D.C.: Cistercian Publications, 1972.
LaFon, Victor. "Histoire de la fondation de l'abbaye de Locdieu." *Mémoires de la société des lettres, sciences et arts de l'Aveyron* 11 (1879): 339–96.
Laporte, J. "Une tentative de fondation cistercienne en forêt de Brotonne." *Revue Mabillon* 44 (1954): 1–5.
Latouche, Robert. "Défrichement et peuplement rural dans la Maine du IXe au XIIIe siècle." *Le Moyen Age* 54 (1948): 77–87.
———. "L'économie agraire et le peuplement des pays bocagers." *Revue de synthése* 17 (1939): 45–50.
Laurent, J. "Les noms des monastères cisterciens dans la toponymie européenne." *Saint Bernard et son temps. Association bourguignonne des sociétés savantes; recueil de mémoires présentés au congrès de 1927*, 1: 168–204. Dijon: Association, 1928–29.
———. "Le problème des commencements de Cîteaux." *Annales de Bourgogne* 6 (1943): 213–29.
Lebecq, Stéphane. "Les cisterciennes de Vaucelles en Flandre maritime au XIIIe siècle." *Revue du Nord* 54 (1972): 371–84.
———. "Vignes et vins de Vaucelles: une esquisse." *L'économie cistercienne. Géographie—Mutations du Moyen Age aux Temps modernes. Publications de la Commission d'histoire de Flaran*, 3 (Auch: Comité départemental du Tourisme du Gers, 1983), pp. 197–206.
Leblanc, Gratien. "La grange cistercienne de Fontcalvi (Aude)." In *Fédération historique du Languedoc Méditerranéen et du Roussillon XXXe et XXXIe Congrès. Montpellier*, pp. 43–57. Sête-Beaucaire: Fédération, 1956–57.
———. "La grange Lassale. Etude historique et archéologique d'une 'grange' cistercienne." In *Xe Congrès de la fédération des sociétés savantes Languedoc-Pyrénées-Gascogne à Montauban, 1954*, pp. 3–16. Montauban: Fédération, 1954.
———. "La répartition géographique des abbayes cisterciennes du sud-ouest." In *France méridionale et Pays ibériques. Mélanges géographiques offerts en hommage à M. Daniel Faucher*, 2: 584–608. Toulouse: Privat, 1949.
LaClercq, Jean. "Comment vivaient les frères convers." *I Laici nella "Societas Christiana" dei secoli XI e XII. Atti della terza Settimana internazionale di studio, Mendola, 21–27 agosto, 1965*, pp. 152–82. Milan: Università Cattolica del Sacro Cuore, 1968.
———. "L'érémitisme du XI siècle dans son contexte économique et social." *Eremitismo in Occidente nei secoli XI e XII. Atti della seconda Settimana internazionale di studio: Mendola, 30 agosto-6 settembre, 1962*, pp. 27–44. Milan: Università Cattolica del Sacro Cuore, 1968.
———. "L'érémitisme et les cisterciens," vol. cit. 573–80.
———. "The Monastic Crisis of the Eleventh and Twelfth Centuries." In *Cluniac Monasticism*

in the Central Middle Ages, ed. Noreen Hunt, pp. 215-37. Hamden, Conn.: Archon Books, 1971.

Lefèvre, Simone. "Un entrepreneur de défrichements au XIIe siècle: David de la Forêt." *Paris et Ile-de-France. Mémoires publiés par la fédération des sociétés historiques et archéologiques de Paris et de l'Ile-de-France* 28 (1978): 77-83.

Le Goff, Jacques and Ruggiero Romana. "Paysages et peuplement en Europe après le XIe siècle." *Etudes rurales* 17 (1965): 5-24.

Lekai, Louis J. *The Cistercians, Ideal and Reality*. Kent, Ohio: Kent State University Press, 1977.

——. "Le collège St. Bernard au moyen âge: (1280-1533)." *Annales du Midi* 85 (1973): 251-60.

——. "Ideals and Reality in Early Cistercian Life and Legislation." In *Cistercian Ideals and Reality*, ed. John R. Sommerfeldt, pp. 4-19. Kalamazoo, Mich.: Cistercian Publications, 1978.

——. *The White Monks*. Okauchee, Wisc.: Our Lady of Spring Bank Press, 1957.

Lenglet, Marie-Odile. "La biographie du Bienheureux Géraud de Salles." *Cîteaux: comm. cist.* 29 (1978): 7-40.

——. "Un problème d'histoire monastique: La fondation de l'abbaye de Mazan." *Cîteaux: comm. cist.* 21 (1970): 5-22.

Le Roy Ladurie, Emmanuel. *The Peasants of Languedoc*, trans. John Day. Chicago: University of Illinois Press, 1974.

——. *Montaillou, village occitan de 1294 à 1324*. Paris: Gallimard, 1975.

Lesne, Emile. *Histoire de la propriété ecclésiastique en France*. 6 vols. Lille-Paris: Faculté catholique de Lille, 1930.

Lewis, Archibald R. "The Closing of the Medieval Frontier, 1250-1350." *Speculum* 33 (1958): 475-83.

——. *The Development of Southern French and Catalan Society: 718-1050*. Austin, Texas: University of Texas Press, 1965.

——. "Patterns of Economic Development in Southern France, 1050-1271 A.D." *Studies in Medieval and Renaissance History*, n.s. 3 (1980): 57-83.

Little, Lester K. *Religious Poverty and the Profit Economy in Medieval Europe*. Ithaca, N.Y.: Cornell University Press, 1978.

Locatelli, R. "La région de Pontarlier au XIIe siècle et la fondation de Mont-Sainte-Marie." *Mémoires de la société pour l'histoire du Droit et des Institutions des pays bourguignons, comtois, et romands* 28 (1967): 1-87.

Lohrmann, Dietrich. "La grange de Troussures-Sainte-Eusoye et le défrichement de la forêt de Noirvaux au XIIe siècle." *Cîteaux: comm. cist.* 26 (1975): 175-84.

——. "Le rétablissement du grand domaine à faire-valoir direct en Beauvais au XIIe siècle." *Francia. Forschungen zur Westeuropäischen Geschichte* 8 (1980): 105-26.

——. "La tuilerie cistercienne de Commelles en forêt de Chantilly." *L'économie cistercienne. Géographie—Mutations du Moyen Age aux Temps modernes. Publications de la Commission d'histoire de Flaran*, 3 (Auch: Comité départemental du Tourisme du Gers, 1983), pp. 213-8.

——. "Répartition et création de nouveaux domaines monastiques au XIIe siècle. Beauvaisis—Soissonnais—Vermandois." *Villa—Curtis—Grangia. Landwirtschaft zwischen Loire und Rhein von der Römerzeit zum Hochmittelalter. Economie Rurale entre Loire et Rhin de l'Epoque Gallo-Romaine au XIIe-XIIIe siècle. Beihefte der Francia*, 11, ed. Walter Janssen and Dietrich Lohrmann, preface by Charles Higounet, pp. 242-59. Munich: Artemis Verlag, 1983.

Lucet, Bernard. *Les codifications cisterciennes de 1237 et de 1257*. Paris: C.N.R.S., 1977.

Lynch, Joseph H. "Cistercians and Underaged Novices." *Cîteaux: comm. cist.* 24 (1973): 283-97.

——. *Simoniacal Entry into Religious Life from 1000 to 1260*. Columbus, Ohio: Ohio State University Press, 1976.

Lyon, Bryce. "Medieval Real Estate Developments and Freedom." *American Historical Review* 63 (1957): 47-61.

Maas, Walther. *Les moines-défricheurs. Etudes sur les transformations du paysage au moyen âge aux confins de la Champagne et de la Lorraine*. Moulins: n. pub., 1944.

Magnou-Nortier, Elisabeth. *La société laïque et l'église dans la province ecclésiastique de Narbonne (zone cispyrénéene) de la fin du VIIIe à la fin du XIe siècle*. Toulouse: Université de Toulouse-le Mirail, 1974.

——. "La terre, la rente et le pouvoir dans les pays de Languedoc pendant le haut Moyen

Age." *Francia. Forschungen zur Westeuropäischen Geschichte* 9 (1981): 79–115; 10 (1982): 21–66.
Mahn, J.-B. *L'ordre cistercien et son gouvernement des origines au milieu du XIIIᵉ siècle (1098–1265)*. Paris: Boccard, 1945.
Marilier, J. "Les débuts de l'abbaye de Cîteaux." *Mémoires de la société pour l'histoire du droit et des institutions des anciens pays bourguignons, comtois, et romands* 15 (1953): 117–22.
Martin-Lorber, Odile. "Une communauté d'habitants dans une seigneurie de Cîteaux aux XIIIᵉ et XIVᵉ siècles." *Annales de Bourgogne* 117 (1958): 7–36.
———. "L'exploitation d'une grange cistercienne à la fin du XIVᵉ siècle et au début du XVᵉ siècle." *Annales de Bourgogne* 115 (1957): 161–80.
McClosky, Donald. "The Economics of Enclosure: A Market Analysis." In *European Peasants and their Markets*, ed. William N. Parker and Eric L. Jones, pp. 123–60. Princeton, N.J.: Princeton University Press, 1975.
———. "The Persistence of English Common Fields." ibid., 73–119.
McCrank, Lawrence J. "The Economic Administration of a Monastic Domain by the Cistercians of Poblet, 1150–1276." In *Studies in Medieval Cistercian History II*, ed. John R. Sommerfeldt, pp. 135–65. Kalamazoo, Mich.: Cistercian Publications, 1976.
———. "The Frontier of the Spanish Reconquest and the Land Acquisitions of the Cistercians of Poblet, 1150–1276." *Analecta cisterciensia* 29 (1973): 57–78.
———. "The Cistercians of Poblet as Landlords: Protection, Litigation, and Violence on the Medieval Catalan Frontier." *Cîteaux: comm. cist.* 28 (1976): 255–83.
McGuire, Brian Patrick. *The Cistercians in Denmark: Their Attitudes, Roles, and Functions in Medieval Society*. Kalamazoo, Mich.: Cistercian Publications, 1982.
Mikkers, Edmund. "Eremitical Life in Western Europe during the XIth and XIIth Centuries." *Cîteaux: comm. cist.* 14 (1963): 44–54.
———. "L'idéal religieux des frères convers dans l'ordre de Cîteaux aux 12ᵉ et 13ᵉ siècles." *Collectanea ordinis cisterciensium reformatorum* (Rome) 24 (1962): 113–29.
Miskimin, Harry A. *Money, Prices, and Foreign Exchange in Fourteenth-century France*. New Haven: Yale University Press, 1963.
Moore, R. I. "St. Bernard's Mission in Languedoc." *Bulletin of the Institute of Historical Research* 47 (1974): 1–14.
Mousnier, Mireille. "Les granges de l'abbaye cistercienne de Grandselve (XIIᵉ–XIVᵉ siècle)." *Annales du Midi* 95 (1983): 7–28.
———. "L'abbaye cistercienne de Grandselve du XIIᵉ au début du XIVᵉ siècle." *Cîteaux: comm. cist.* 34 (1983): 53–76, 221–44.
Mundy, John Hine. "Charity and Social Work in Toulouse, 1100–1250." *Traditio* 22 (1966): 203–87.
———. *Liberty and Political Power in Toulouse, 1050–1230*. New York: Columbia University Press, 1954.
———. "Urban Society and Culture: Toulouse and its Region." In *Renaissance and Renewal in the Twelfth Century*, ed. Robert L. Benson and Giles Constable, pp. 229–47. Cambridge, Mass.: Harvard University Press, 1982.
———. "Village, Town, and City in the Region of Toulouse." In *Pathways to Medieval Peasants. Papers in Medieval Studies II*, ed. J. Ambrose Raftis, pp. 141–90. Toronto: Pontifical Institute of Mediaeval Studies, 1981.
Niermeyer, J.-F. *Mediae latinitatis lexicon minus*. Leiden: E. J. Brill, 1974.
Noël, Raymond. *Dictionnaire des châteaux de l'Aveyron*. 2 vols. Rodez: Subervie, 1971.
Noël, René. "Orval et l'économie cistercienne aux XIIᵉ et XIIIᵉ siècles." In *Aureavallis: Mélanges historiques réunis à l'occasion du neuvième centenaire de l'abbaye d'Orval. Colloque, 1970*, pp. 283–91. Liège: Soledi, 1970.
Noblet, A. "Chronique bibliographique: ordre de Cîteaux." *Revue Mabillon* 2 (1906): 83–93.
North, Douglass C. and Robert Paul Thomas. *The Rise of the Western World: A New Economic History*. Cambridge: Cambridge University Press, 1973.
Omont, H. "La Collection Doat à la Bibliothèque Nationale: Documents sur les recherches de Doat dans les archives du sud-ouest de la France de 1663 à 1670." *Bibliothèque de l'Ecole des Chartes* 77 (1916): 286–336.
Othon, R. P. "De l'institution et des convers dans l'ordre de Cîteaux (XIIᵉ et XIIIᵉ siècles)." In *Saint Bernard et son Temps. Recueil des mémoires et communications présentés au congrès*, 2: 139–201. Dijon: L'Académie des sciences, arts et belles-lettres de Dijon, 1929.

Ourliac, Paul. "Les villages de la région toulousaine au XIIe siècle." *Annales: E.S.C.* 4 (1949): 268–77.
———. "La 'convenientia.'" In *Etudes d'histoire du droit privé offertes à P. Petot*, pp. 413–22. Paris: Montchrestien, 1959.
———. "Le pays de la Selve à la fin du XIIe siècle." In *Les structures sociales de l'Aquitaine, du Languedoc, et de l'Espagne au premier âge féodal*. Colloques Internationaux du Centre National de la Recherche Scientifique: Toulouse, 28–31 mars, 1968, pp. 239–60. Paris: C.N.R.S., 1969.
———. "Réflexions sur le servage languedoçien." *Comptes rendus de l'Académie des inscriptions et belles lettres* 98 (1971): 585–91.
Petri, Franz. "Entstehung und Verbreitung der Niederländischen Marschen Kolonisation in Europa (mit Ausnahme der Ostsiedlung)." In *Die Deutsche Ostsiedlung des Mittelalters als problem der Europäische Geschichte, Reichenau: Vorträge. 1970-72*, ed. Walter Schlesinger, pp. 695–754. Sigmaringen: Thorbecke, 1975.
Pfleger, L. "Leibrentenvertraje in Zisterzienserklostern." *Cistercienserchronik* 17 (1905): 118–20.
Plaisance, G. "Les cisterciens et la forêt." *Revue du Bois* 10 (1955): 3–8.
Platt, Colin. *The Monastic Grange in Medieval England: A Reassessment*. London: Macmillan, 1969.
Poloni, Jacques. "Les granges de l'abbaye de Cîteaux." *L'économie cistercienne. Géographie—Mutations du Moyen Age aux Temps modernes. Publications de la Commission d'histoire de Flaran*, 3 (Auch: Comité départemental du Tourisme du Gers, 1983), pp. 183–88.
Poly, Jean-Pierre. *La Provence et la société féodale, 879-1166. Contribution à l'étude des structures dites féodales dans le Midi*. Paris: Bordas, 1976.
Pontus, Paul. *Silvacane Abbey*. Paris: Caisse Nationale des Monuments Historiques, 1966.
Portela Silva, Ermelindo. *La colonizacion cisterciense en Galicia (1142-1250)*. Santiago de Compostela: Universidad de Santiagó, 1981.
Power, Eileen. *The Wool Trade in English Medieval History*. Oxford: Oxford University Press, 1941.
Raftis, J. Ambrose. "Western Monasticism and Economic Organization." *Comparative Studies in Society and History* 3 (1960–61): 452–69.
Raindorf-Gérard, Christiane. "Les origines de l'abbaye cistercienne de Lieu-Saint-Bernard (1237–1246)." In *Hommage au Professeur Bonenfant: 1899-1965. Etudes d'histoire médiévale dédiées à sa mémoire par les anciens élèves de son seminaire à l'université libre de Bruxelles*, pp. 197–208. Brussels: G. Despy, 1965.
Regne, Jean. "L'abbaye de Mazan de 1123 à 1500." *Revue Mabillon* 14 (1924): 214–31.
Reyerson, Kathryn L. "Le rôle de Montpellier dans le commerce des draps de laine avant 1350." *Annales du Midi* 94 (1982): 17–40.
Ribbe, Wolfgang. "Die Wirtschaftstätigkeit der Zisterzienser im Mittelalter: Agrarwirtschaft." *Die Zisterziener. Ordensleben zwischen Ideal und Wirklichkeit. Eine Ausstellung des Landschaftsverbandes Rheinland, Rheinisches Museumsamt, Brauwiler. Aachen, Krönungssaall des Rathauses 3. Juli-28. September 1980*, pp. 203–215. Cologne: Rheinland-Verlag, 1980.
Richard, Jean. "Les débuts de la Bussière et de Fontenay." *Mémoires de la société pour l'histoire du droit et des institutions des anciens pays bourguignons, comtois et romands* 15 (1953): 123–29.
Richardot, Hubert. "Le fief roturier à Toulouse aux XIIe et XIIIe siècles." *Revue historique du droit* 40 (1935): 307–59.
Robert, A. "Les abbés du monastère cistercien des Chambons au diocèse de Viviers (1152–1791)." *Revue de Vivarais* 72 (1968): 119–30; 73 (1969): 30–35.
Roblin, Michel. "L'habitat rural dans la vallée de la Garonne de Boussens à Grenade." *Revue géographique des Pyrénées et du Sud-Ouest* 8 (1937): 5–72.
Roehl, Richard. "Plan and Reality in a Medieval Monastic Economy: the Cistercians." *Studies in Medieval and Renaissance History* 9 (1972): 83–113.
Rösener, Werner. "Bauernlegen durch Klösterliche Grundherren im Hochmittelalter." *Zeitschrift für Agrargeschichte und Agrarsoziologie* 27 (1979): 60–93.
———. "L'économie cistercienne de l'Allemagne occidentale (XIIe–XVe siècle)." *L'économie cistercienne. Géographie—Mutations du Moyen Age aux Temps modernes. Publications de la Commission d'histoire de Flaran*, 3 (Auch: Comité départemental du Tourisme du Gers, 1983), pp. 135–56.
Rumeau, M. R. "Notes sur l'abbaye de Grandselve." *Bulletin de la société de géographie de Toulouse* 19 (1900): 247–85.

Saint-Aubin, A.-M. de. "La fondation de l'abbaye de Bellevaux." *Mémoires de la société pour l'histoire du droit et des institutions des anciens pays bourguignons, comtois, et romands* 15 (1953): 178–80.

Saint-Blanquat, Odon de. "Comment se sont créées les bastides du sud-ouest de la France." *Annales* 4 (1949): 278–89.

Saint-Jean, Robert. "L'abbaye cistercienne de Mazan (Ardèche) et ses filles provençales: Sénanque et le Thoronet." *Fédération historique du Languedoc Méditerranéen et du Roussillon 90ᵉ congrès* (1968): 77–100.

Salmon, J. "Morimond et ses granges." In *Actes du 38ᵉ Congrès de l'association bourguignonne des sociétés savantes*, pp. 105–25. Langres: Société historique et archéologique, 1969.

Santacana Tort, Jaime. *El monasterio de Poblet (1151–1181)*. Barcelona: Consejo Superior de Investigationes Científicas, 1974.

Sclafert, Thérèse."A propos du déboisement des Alpes du Sud." *Annales de géographie* 42 (1933): 266–77.

———. *Cultures en Haute-Provence: Déboisements et pâturages au Moyen Age*. Paris: S.E.V.P.E.N., 1953.

Schlesinger, Walter. "Flemmingen und Kühren zur Siedlungsform niederländischer Siedlungen des 12. Jahrhunderts im Mitteldeutschen Osten." In *Die Deutsche Ostsiedlung des Mittelalters als problem der Europäische Geschichte, Reichenau: Vorträge, 1970–72*, ed. Walter Schlesinger, pp. 263–309. Sigmaringen: Thorbecke, 1975.

Slicher van Bath, B. H. *The Agrarian History of Western Europe, A.D. 500–1850*, trans. Olive Ordish. London: E. Arnold, 1963.

Sorre, Max. "Etude sur la transhumance dans la région montpellieraine." *Bulletin de la Société languedocienne de géographie* 35 (1912): 1–40.

Southern, R. W. *Western Society and the Church in the Middle Ages*. Harmondsworth: Penguin, 1970.

Tardieu, Joëlle. "Le cellier de l'abbaye de Léoncel à Beaufort-sur-Gervanne." *Cîteaux dans la Drôme: Revue Drômoise* 83 (1980): 99–106.

Taupiac, Louis. "L'abbaye de Belleperche." *Bulletin de la société archéologique de Tarn-et-Garonne* 6 (1878): 89–115.

Tellenbach, Gerd. "Il monachesimo riformato ed i laici nei secoli XIᵉ XII." *I Laici nella "Societas Christiana" dei secoli XI e XII*, Atti della terza Settimana internazionale di studio: Mendola, 21–27 agosto, 1965, pp. 118–51. Milan: Università Cattolica del Sacro Cuore, 1968.

Thiele, Augustinus. *Echternach und Himmerod. Beispiele benediktinischer und zisterziensischer Wirtschaftsfuhrung im 12. und 13. Jahrhundert*. Stuttgart: Gustav Fischer Verlag, 1964.

Thompson, Sally. "The Problem of Cistercian Women in the Twelfth and Thirteenth Centuries." In *Medieval Women, Studies in Church History: Subsidia I*, ed. Derek Baker, pp. 227–52. Oxford: Blackwell, 1978.

Thouzellier, Christine. *Hérésie et hérétiques. Vaudois, Cathares, Patarins, Albigeois*. Rome: Edizioni di Storia et Letteratura, 1969.

Toubert, Pierre. *Les structures de Latium médiévale: Le Latium méridional et la Sabine du IXᵉ siècle à la fin du XIIᵉ siècle*, 2 vols. Rome: Ecole Française, 1973.

Traissac, Elisabeth. "Les abbayes cisterciennes de Fontguilhem et du Rivet et leur rôle dans le défrichement médiévale en Bazadais." *Revue historique de Bordeaux* 9 (1960): 141–50.

Trilhe, Robert. "A propos de la fondation de l'abbaye d'Eaunes." *Revue de Comminges* 28 (1915): 253–62.

Vann Damme, Jean-Baptiste. "Moines—Chanoines—Cîteaux. Influences réciproques." In *Aureavallis. Mélanges historiques réunis à l'occasion du neuvième centenaire de l'abbaye d'Orval: Colloque, 1970*. Liège: Soledi, 1970.

Van der Meer, F. *Atlas de l'Ordre cistercien*. Paris: Editions Sequoia, 1965.

Verna, Catherine. "La sidérurgie cistercienne en Champagne méridionale et en Bourgogne du Nord (XIIᵉ–XVᵉ siècle)." *L'économie cistercienne. Géographie—Mutations du Moyen Age aux Temps modernes*. Publications de la Commission d'histoire de Flaran, 3 (Auch: Comité départemental du Tourisme du Gers, 1983), pp. 207–12.

Viard, Pierre-Paul. "Saint Bernard et les moines décimateurs." In *Saint Bernard et son Temps. Recueil des mémoires et communications présentés au congrès*, 2: 292–94. Dijon: Académie des sciences, arts et belles-lettres de Dijon, 1929.

Wakefield, Walter L. *Heresy, Crusade and Inquisition in Southern France. 1100–1250*. Berkeley: University of California Press, 1974.

Weiss, Hildegard. *Die Zisterzienserabtei Ebrach. Eine Untersuchung zur Grundherrschaft, Gerichtherrschaft und Dorfgemeinde im frankischen Raum*. Stuttgart: Gustav Fischer Verlag, 1962.

White, K. D. *Roman Farming*. Ithaca, N.Y.: Cornell University Press, 1970.
White, Lynn, Jr. *Medieval Technology and Social Change*. New York: Oxford, 1962.
Wiswe, Hans. "Grangien niedersächsischer Zisterzienserklöster. Entstehung und Bewirtschaftung spätmittelalterlich-frühneuzeitlicher landwirtschaftlicher Grossbetriebe." *Braunschweigisches Jahrbuch* 34 (1953): 21–134.
Wolff, Philippe. *Commerces et marchands de Toulouse (vers 1350–vers 1450)*. Paris: Plon, 1954.
———. "Villes et campagnes dans le Midi français médiéval." *France méridionale et Pays ibériques. Mélanges géographiques offerts en hommage à M. Daniel Faucher*, 2: 677–85. Toulouse: Privat, 1949.
Young, Charles R. *The Royal Forest of Medieval England*. Philadelphia: University of Pennsylvania Press, 1979.

MAPS CONSULTED

Cartes de France, ed. César François Cassini de Thury (1714–84).
Cartes touristiques, 1:100,000, ed. Institut Géographique National de France.
Cartes de France, 1:50,000, ed. Institut Géographique National de France.

INDEX

Abbeys, sites for, 11, 29–30
Abreuvage, watering rights, 109n68
Ademar, abbot of Bonneval, 113n85
Ademar, bishop of Rodez, 65n21
Ademar of Cairo, 20
Agen, 82–83, 123
Agricultural practice:
   adaptation to local conditions, 5, 10, 57, 93, 118, 124–27
   advantages of Cistercian grange, 30, 74–93
   crop diversity, 70–71
   direct management, 61, 76–78, 84
   lands rested to restore fertility, 76–77
Aicelina, widow of Raymond Talairan, 38n36
Aiguebelle, abbey of:
   folding sheep, 100n33;
   foundation, 3n8;
   pasture rights, 106
Aiguebelle, land at, 25n54, 116n104
Aigues-Mortes, 24
Aimeric of St. Paragoire, 44
Albarelle, territory of, 22
Albert, abbot of Berdoues, 16n15
Albi, bishops of, 3
Albi, commercial ties to, 68n37, 123
Albigensian Crusade, 33n17, 64n15, 124
Albigeois, 15, 105, 108, 111
Alexander III, 67
Alexander IV, 70
Aliena animalia (outsiders' animals), 115–16
Allod, allodial rights, 13, 20–21
Alps. See pastoralism—locations, or—summer pasture.
Amblard, abbot of Bonnecombe, 109n68
Ampiac, tithes and church of, 52, 64n74, 65
Anastais, 25n50
Andreas Delphinus, count of Vienne, 105n50
Animals and animal husbandry. See pastoralism and transhumance
Annexation of earlier religious houses, 31–34
Annual rents paid, 37, 51
Anti-Cistercian polemic, 5n17
Apas, grange of Bonnefont, 63
Appendaria, appendariae, 13, 15, 45
Aragon, king of, Alphonse, 82
Arbert of Turre, 96n8
Arbert, abbot of Locdieu, 113n85
Arbitrated settlements. See *convenientiae*
Arcambald of Panat, 35n26, 101n35
Ardichol, territory of, 48
Ardorel: abbey of:
   finances, 118

   foundation, 32
   pasture rights, 111
   site change, 73n65
Ariège, region of, 28, 107–11
Armagnac, lords of, 107
Arnold, abbot of Berdoues, 13n7
Arnold Amalric, abbot of Cîteaux, earlier of Grandselve, 64n15, 129n42
Arnold Helye, 45–46
Arnold of Oms, 14
Arnold of St. Roman, 99n27
Arnold of Taurinus, 13n5
Arnold William of Lobcascha, 49n78
Artigues, grange of Berdoues, 29–30, 32, 34n21, n23, 62n4, 88n96
Artigues, place-names related to, 30. See also assarts
Ascent and descent of animals. See pastoralism and transhumance
Assarts, by peasants, 125n36
Assarts, tendency of order to acquire, 29–30, 58, 98n17
Assou, grange of Fontfroide, 67n33
Ato of Cums, 13
Aubrac plateau, 72, 112–113, 115
Aubrac, hospital of, 63n12
Auriac, parish of, 71
Austorga, 76n71
Aveyron River, 71, 113–114

Bagnols, grange of Grandselve, 45–46, 69n41
Bannal of seigneurial rights, 89–91
Bar, grange of Bonnecombe, 71
Barcelona, Raymond, count of, 82
Bargas, *mansus* and *caputmansus* of, 21n37
Barns. See granges
Basques, 65n16
Bastides, 63, 70, 120
Béarn, counts of, 34, 107
Beaucaire, grange of Bonneval, 67
Beaucaire, city of, 115
Becerières, *appendaria* of, 13
Beceria, parish of St. Peter of, 17
Bedored, grange of Berdoues, 43n54, 63n11
Bego Ysarnz of Mirabelle, 95n2
Beliarda of Comas, 16n15
Belleperche: abbey of:
   churches, 34n22
   disputes, 55
   documents, 2, 31
   foundation, 2
   granges, 66n26
   pastoralism, 107–108

Belloc, abbey of:
  documents, 2n3
Belvezins, parish of, 25n50
Benedictine Rule, 1, 100
Benedictines, 121, 124, 126
Beneficium, 19
Bequests. See gifts
Berdoues, abbey of:
  abbots: Albert, 16n15; Arnold, 13n7; Hugh, 99n26; Vidal, 108n63; William, 57n114, 20n33, 14n8
  cartulary, 2, 16, 106
  casaux, 16, 65n17
  clearance, 28, 48n74
  commercial interests, 122
  conveyances by: Arnold of St. Roman, 99n27; Arnold of Oms, 14; Arnold William of Lobcascha, 49n78; Beliarda of Comas, 16n15; Bernard of Mesplede, 16n15; Bernard, count of Astarac, 36n28; Bonetus of Lasmeades, 47; Bruno of Lacome, 20; Gassiarnaldus of Marrast, 13n7; Gualardus, son of Jourdain of Marrast, 13n7; Guillelmus Gassias of Orbezan, 42n49; Jourdain of Marrast, 13; Peter of Orbezan, 42n49; Raimundus Hospitale, 48n74; Roger, nephew of Peter of Serres, 42n50; Sanche Lub of Faisan, 42n50; Sanche, son of Bernard of Astarac, 36n28; William of Laseran, 46; William Sanz de Lisos, 48n74
  forest rights, 98
  foundation, 2, 32, 36n28
  granges described, 65, 69
  granges; Artigues, 29–30, 32, 34n21, n23, 62n4, 88n96; Bedored, 43n54, 63n11; Cuelas, 48; Esparciac, 48; St. Elix, 62n4
  hermits and hermitages, 30, 32, 34n23
  mills, 88n96, n99
  mortgages, 37n30
  murder at, 58–59
  pastoralism and transhumance, 99n27, 106–107
  recruiting, 48n78, 78
  site, 2, 32
  ties to l'Escaledieu, 107n60
  tithes, 42n51
Berdoues, land, church, and *casale*, 36n28
Berengairenc, *mansus* of, 13, 76n71
Berengar, archbishop of Narbonne, 102n39
Berengar of Auriac, 35n26
Berengar Rotbald of Capestang, 102
Bernac, grange of Bonnecombe, 20, 57n114, 65, 68n37–39, 71, 114
Bernard, abbot of Grandselve, 87n95
Bernard, abbot of Gimont, 28n65, 47n70
Bernard, archbishop of Auch, 49n78
Bernard, count of Astarac, 36n28, 42n49
Bernard Carpenter, 20
Bernard Ermengaudi, 27n58
Bernard Gallard, 35n29

Bernard of Arpaian, 21n37
Bernard of Bez, 19, 35n25
Bernard of Capraria, 27n58
Bernard of Clairvaux, 1–5, 32, 124n30, 128
Bernard of la Roca, 21n37
Bernard of las Bordas, 19
Bernard of Levezou, 101
Bernard of Maselz, 20
Bernard of Mesplede, 16n15
Bernard of Montaut, 47
Bernard of St. Germain, 19
Bernard Pilgrim, 46
Bernard Terz, 38n36
Bertrand, abbot of Grandselve, 45n65
Bertrand, abbot of Bonnecombe, 36n29
Bertrand of Balaguier, 101
Bertrand of Marrencs, 34n23
Bertrand of Ponte, 96n7
Bescabes, forest of, 98n19
Bez, *mansus* of, 19, 35n25, 36n29
Béziers, 51, 111, 124
Biac, grange of Bonneval, 67, 72
Bion, territory of, 67n29
Bodigas (brambles), 27n59
Bogath, field at, 57n114
Bolhaguet, grange of Grandselve, 69n41
Bonauberc, grange of Bonneval, 72
Bonetus of Lasmeades, 47
Bonnecombe, abbey of:
  abbots: Amblardus, 109n68; Bertrand, 36n29; P. 91n111; Pons, 13n6; Randulf, 19n27; Ugo, 21n37; Willermus, 95n2
  administrators, 35n26
  cartularies and documents, 3, 20, 43n53, 64, 114n95
  commercial ties, 69n44, 114, 123
  confirmations, 68, 100n29
  consolidation, 21
  conveyances by: Ademar of Cairo, 20; Aicelina, widow of Raymond Talairan, 38n36; Arcambald of Panat, 35n26; Arnold of Taurinus, 13n5; Ato of Cums, 13; Austorga, 76n71; Bego Ysarnz of Mirabelle, 95n2; Berengar of Auriac, 35n26; Bernard Carpenter, 20; Bernard of Arpaiano, 21n37; Bernard of Bez & son Bernard, 19; Bernard of la Roca, 21n37; Bernard of Maselz, 20; lords of Calmont, 23n44; Déodat of Caylus, 35n26; Frotard of Turre, 109n68; G. of la Roca, 21n37; Gago of Peyrebrune, 35n26; Guillelma, 109n68; Hector of Mirabel, 35n26; Hugh of Gaillac, 20; Lambert of Castelpers, 35n26; Matfres, 76n71; Montarsin of Calmont, 35n26; Panat family, 35n26, 37n30; Peter Carpenter, 20; Peter of Teillz, 21n37; Peter of Turre, 35n26; Peter Ugo, 76n71; Peter Ymbert, 35n26, 38n35; Petronilla, wife of Bernard of la Roca, 21n37; Poncia, sister of Arnold of Taurinus, 13n5; Ray-

mond of Castronovo, 35n26; Rigaldus Amblartz, 21n37; counts of Rodez, 23n44; Roterius of Solis, 35n26; counts of Toulouse, 23n44; Ugo Bonefons, 35n25; Ugo of la Roca, 35n26; Ugo of Seveirac, 36n29; Ugo, count of Rodez, 21n37; William Seinnorellz, 50n86
cultivated land acquired, 15, 19–21
direct management, 77–78
disputes, 3, 86n92, 101, 112, 127
expenditures, 38
foundation, 3, 35n26
granges, 64, 65, 66n26, 70–71
granges: Bernac, 20, 57n114, 65, 68n37–39, 114; Bonnefon, 63n11, 68n37–39, 71–72, 103, 114; Bougaunes, 24, 62, 68n37, 71, 93n117; Iz, 44n56, 52, 64, 65, 68n36–39, 71–72; Magrin or Lafon, 15n11, 20–21, 44n56, 68n36–39, 70; Mansedieu, 68, 113n88; Moncan, 68n36–39, 70–71; St. Félix, 62n4, 64, 68n38–39, 71–72; Vareilles, 13, 56, 68n36–39, 70
hospice, 50n86
location, 3
mills and milling rights, 68n38, 87n95, 88n98, n99
mortgages, 39n41
other holdings, 20, 68n37, n38, 71
pastoralism and transhumance, 70–71, 113–115
recruiting, 56, 78
tithes and churches, 51–52, 65n20, 68n38
urban properties, 68n37, n38, 123
villagers, rights over, 20, 91n111
vineyards and viticulture, production of wine, 24, 71, 92–93
wealth, 64
Bonnefon, grange of Bonnecombe, 63n7, 68n37–39, 71–72, 103, 114. See also la Serre
Bonnefont, abbey of, 41n48
confirmations, 70
documents, 2
foundation, 2, 3n8, 32, 58
granges described, 70
granges: Apas, 63; Minhac, 70, 72
mills, 88n99
pasture rights, 106–10
salt, 102
site, 30
Bonneval, abbey of: (in the Rouergue, not to be confused with Bonnevaux, east of the Rhône)
abbots: Ademar, 113n85; Guillelmus, 113n85
annexations, 112–13
commercial ties, 122–123
confirmations, 67
disputes, 112–15
documents, 3, 31n1, 114n95

filiation, ties, 115
foundation, 3, 86n92, 113
granges, 66n26, 71–72
granges: Beaucaire, 67; Biac, 67, 72; Bonauberc, 72; Empiac, 67; La Fraissenet, 67; La Roquette, 67, 72, 114; La Serre, 38n36, 67, 71–72, 103, 114; La Vayssière, 72; Les Gallinières, 62, 63n12, 72; Masse, 67, 72; Montégut, 67; Paulette, 67; Pomers, 67; Previnquières, 67; Puszac, 67, 72; Séveyrac, 72; Tegula, 67; Verruca, 67
pastoralism and transhumance, 112–115
site, 3, 71–72
tithes and churches, 17–18, 52
wealth, 3
Bonnevaux, abbey of: (in the Marquisate of Provence; not to be confused with Bonneval in the Rouergue)
abbots: John, 105n50
commercial ties, 4, 69n44
conveyances by: Andreas Delphinus, count of Vienne, 105n50; Guigo, count of Vienne, 105n50; Ismio Ruvoiria, 25n54; Matilda, countess of Vienne, 105n50; Simon of Elbone, 116n104
foundation, 3–4, 36
irrigation, 26, 90
meadows and uprooted areas, 26
mills, 88n99
pasture rights, 105
site, 4
Bonus of Perissan, 47n70
Boralde ravine, 114
Bordaria, defined, 44–45
Bordeaux, 82, 93, 102
Bordelz, mansus of, 21n37
Borrowing. See mortgages
Boseville, casal and territory of, 22, 38n36
Bougaunes, grange and vineyards of Bonnecombe, 24, 62, 68n37, 71, 93n117
Bouillas, abbey of:
pasture rights, 110
Boulbonne, abbey of:
annexing smaller religious houses, 29n66, 33
clearance, 28–29
confirmation, 33n17
disputes, 33, 41n48, 100
documents, 2
forest of, 29n66
foundation, 2, 27, 29, 32
granges described, 66n26
granges: Vajal, 29n66, 33, 34n21; Tramesaigues, 33, 34n21, 73n65
hospice, 123
mills, 88n99
pastoralism and transhumance, 109, 110
passage rights, 122
salt, 102
site, 28, 32, 73n65

168   INDEX

Boulhac, church of St. Sulpice of, 51n89
Breeding, animal. See pastoralism and transhumance,
Bremund of Uzès, 111n78
Bremund William of Sommières, 25n50
Brin, *villa* of, 20
Brugale, land at, 47
Bruno of Lacome, 20
Building materials, 97n16
Bulls, 103
Bumper Crops, 8
Burau, forest of, 26, 98
Burau, grange of Valmagne, 26, 66n17, 69, 73
Burial promised to donor, 42n50

Cabanac, mill at, 89n101
Cabanes (shepherds' huts), 98
Cadouin, abbey of, 32n9
Cahors, 82-83
Cairaguet, *mansus* of, 44n56, 59n121
Cairelencz, *mansus* of, 50n86
Cairo, *mansus* of, 20, 44n56, 59n121
Calcantus, forest of, 98
Calcassac, grange of Grandselve, 69n41
Calers: abbey of:
  consolidation, 66
  documents, 2
  foundation, 2
  passage rights, 122
  pasture rights, 109-110
Calixtus II, 36
Calm della Garriga, 68n37
Calm Median, 67n29
Calmels, 95
Calmont, lords of, 23n44
Calviac, lordship of, 71
Camarès, parish of, 111n79, 127n40
Camargue, 24, 111, 114, 122, 125
Camboulas, mills at, 88n98
Campagnes, grange of Villelongue, 34n21, 62n, 73
Campan, 106
Campuslongus, pasture in, 96n7
Canavellas, territory of, 47
Candeil, abbey of:
  abbot Gausbert, 13n5, 35n26, 76n71
  church rebuilt, 34n24, 124
  documents, 3
  forest rights, 98
  foundation, 2-3, 118n1
  mills, 88n99, 91n111
  passage rights, 121
  pasture rights, 111
  urban properties, 123
Canemals, grange of Fontfroide, 67n33
Canon law, 51n92
Cantalops, pasture in, 96n7
Canterbury, archbishop of, 85n90
Canvern, grange of Valmagne, 69, 73

Capéstang, salt at, 102n39
Capital. See cash.
Capriliis, grange of Valmagne, 69
Caputmansus, *caputmansi*, 13, 15
Carabaza, *mansus* of, 35n25, 36n29
Carthusians, 105
Cartularies. See also charters, documents, and entries under individual abbeys. 31, 41, 43, 49, 68, 82-83, 99n25, 126
Casal, casaux, casalium, casalia, 15-16, 64-65
Casamaurel, 68n38
Cash, 36-40, 45, 58, 86-87, 94, 102-03
Cassagnes, property of, 68n38
Castelsarrazin, 107-08
Castres, 123
Catalonia, 4, 73, 122
Cattle, 95, 96
Caucill, *mansi* and *appendaria* of, 35n25, 36n29
Causse Comtal, 71, 72, 114
Causse de Larzac, 95, 111-113
Causse de Sévérac, 114
Cavaillon, 4
Cecilia of Provence, 4, 34n24
Celsan, parish of St. John of, 170
Censuales, 13
Centullus, count of Astarac, 42n49
Cereal consumption, 96
Cereal production:
  advantages of Cistercian grange for, 41, 61, 64, 72-73, 75, 76-93, 94
  seed/yield ratios, 80-81, 84-85, 92
  source of cash, 40, 117
  types produced, 70-71
Cévennes, 105, 110-115
Chalmensus, 105n50
Chalvas, 25n54, 26
Champdesvignes, grange of Valmagne, 69
Charcaleves, territory of, 67n29
Charcoal-making, 98
Charters. See also cartularies, contracts, conveyances, documents
  descriptions of land, 12-16
  division *per alphabetum*, 42n49
  perambulations listed, 44
  permitting clearance, 25-29
  phraseology, 12-16
  produced by monks as evidence, 42
  recruiting, 10n49, 53, 55-57, 61, 79
  vagueness in, 15, 25
Chevremorte, forest of, 98
Choir monks working alongside *conversi*, 41
Chronicles. See Silvanès and Foundation accounts
Church dedications, 17
Churches and tithes, acquired, 16-17, 51n89. See also under individual abbeys
Churches at granges, 62
Cistercian order: See also individual abbeys
  acquisitions, 23
  architecture, 3-4, 7, 41n46, 119n5

INDEX

ascetic life, economic consequences of: 1, 40–41, 83, 97, 120
compared to Cluniacs, Benedictines, 35
debate with Cluniacs, 5
economic model, 1, 5–10
elsewhere, 9n28, n29, 56
exemption from tithes. See tithes, exemption
expansion, 1, 39
foundations in southern France, 1–4, 7n25, 32–36. See also individual abbeys
idealization, 7, 9, 33n20, 127
innovation, 128
legislation. See Statutes
nuns, x, 24n49, 94
contribution to medieval economic growth, 7–9
prehistory, 33–34
reputation for pioneering, 8–10, 97
simplicity, 119–20
sites, 6
success, 1, 5, 10, 124–29
tax-rolls, 69. See also Appendix I.
wealth, 8–10
Citation style, xiv
Cîteaux, abbey of: filiation, 69; foundation, 9
Clairvaux, abbey of,
abbot, 107. See also Bernard of Clairvaux
filiation, 2, 4, 64, 69, 91
Clearance: See also assarts
by Cistercians, 12, 24–25, 27, 97–98
by others in southern France, 9, 24–25, 29–30, 125
Cluniacs, 5, 7, 120n6
Coastal marshes, 25
Cocone, land of, 39n40
Cognerio, grange of Léoncel, 67
Colleges, 123
Coloni, peasants, 18
Colons, parish of St. Martin of, 17
Comas, land of, 16n15
Combacalida, grange of Léoncel, 67
Combelongue, canons of, 41n48, 109–10
Comberouger, grange of Grandselve, 69n41
Commercialization, 123–124
Comminges, 70, 106–07
Compagnes, hermitage and granges of, 32
Complant or *méplant* agreements, 116
Comps, *villa* of, 68n37, 91n111
Condamina, 27n58
Condom, city of, 82–83
Confraternity in benefits of monastery, 13, 38n36. See also religious benefits
Conques, abbey of, 39
Consolidation: 43–50
compacting to create large holdings, 43–48
completeness of, 59–60
patience required, 23, 31
reassembling of rights, 21–22, 46–50
seeking coterminous holdings, 43–50
transformation of rural economy, 40

Contracts, 36–40, 87n95, 94–95, 107–110, 114–116, See charters, cartularies, documents
Convenientia, (*-ae*), 94–95, 107–110, 114
Conversi, laybrothers:
status, 54
diet and productivity, 81
importance to order's success, 54–57, 78–82, 93, 125
recruitment, 19, 49, 53, 55–57, 61, 79
replace tenant farmers, 41, 61, 78
revolts, 56–57, 59
Conveyances to order, 36–40
Corbières, hills of, 73
Corronzanges, parish of, 18, 50, 52
Coubirac, grange of Grandselve, 69n41
Council, Fourth Lateran, 52, 86
Critics of Cistercians, 7
Crop rotation, 91–92
Crops, new or industrial, 76
Crusades, donations upon departure for, 35n26
Cuelas, grange of Berdoues, 48
Cult of relics, 39
Culturae, field strips, 48
Cultus et incultus, 14–15, 27

Dairese, *mansus* of, 38n36
Dalon, abbey of, 112–113
Damage done by animals, 101
Debt, 112–113. See also mortgages
Decadence, 7n25, 17, 52–53, 90–91, 124–25, 129
Decimaria, tithe-producing area, 16
Dedications of churches, 17
Demand for food, clothing, shelter, 96, 117
Demographic growth, 96, 119, 129
Déodat Jovus, 35n25
Déodat of Caylus, 35n26
Dependency cost, hypothetical example, 79–81
Dependent tenants, 12, 43, 53–59. See also peasants
Deusaiuda of Lombirag, 47n70
Development. See improvements
Die, diocese of, 98n17
Diet, 40, 76, 96, 127
Direct management, 61, 76–78, 84
Discipline, internal, 104
Disguised sales, 36
Disincentives to investment, 83
Distribution of personnel, 34
Diversity of grange types, 72–74
Doat Collection, 3
Documents, 2, 31, 34n24, 36n28. See also charters, cartularies, contracts
Dominium. See lordship
Donors and founders, 34–35
Dourdou River, 49
Drainage, 26–27

Drayas, drailles (sheep roads), 105, 114
Durban, grange of Fontfroide, 17, 67n33
Durenque, parish of, 71
Durfort, *castellum*, land, and church of St. Stephen of, 20, 47n78, 48–49

Economic trends favoring stock-raising, 96–97
Efficiencies of Cistercian management, 74–86
Egez, vineyards of, 68n36–38
Eleanor of Aquitaine, donation by, 118n1
Elne, holdings in diocese of, 24, 74
Empiac, grange of Bonneval, 67
Enfranchisement, 19–20
England, kings of, 3n8, 32, 82, 102
Englesa, 77n75
Escaledieu, abbey of:
 founders, 34n24
 granges, 73
 site, 106
 pasture rights, 106–07
 ties to Berdoues, 107n60
Esparciac, *villa* and grange of, 48
Estives, 106, 109–10, 116. See also pastoralism and transhumance
Exchange rates, 37n34
Exempt status of monks, 82
Exheremare, to clear land, 28
Expansion, limits of, 125
Expenditures, 45

Fairs, buying and selling animals at, 104
Fallow land, 27
Felgairetas, *mansus* of, 21n39
Fevum, 13, 19, 46
Filiation within the order, effects of, 112–113, 115, 123
Finage, 29–30, 58. See also assart
Firewood, 97
Fishing rights, 82
Fishponds and fishing rights, 49, 82, 90
Flaran, abbey of:
 church, 119n
 pasture rights, 110
Fleeces and sheepskins, 95n4
Flooding, 30, 72, 74
Foix, 96, 107, 111
Folhac, honor of, 108n63
Fondouce, grange of Valmagne, 69
Font Marcel, territory of, 66n22
Fontfroide, abbey of:
 commercial ties, 69n44
 confirmations, 67
 daughter-abbeys in Spain, 33
 donors: counts of Foix, 111
 granges described, 66n26, n33, 67, 73
 granges: Assou, 67n33; Canemals, 67n33; Durban, 17, 67n33; Livières, 67n33; Pardines, 67n33; Poujouls, 67n33; St. Victor of Montveyre, 67n33; Haulterive, 67n33; Montlaurès, 17, 67n33; Montrédon, 67n33; Salavert, 67n33; St. Martin, 67n33; Ste. Eugenie, 67n33; Taura 67n33
 incorporation, 3, 32
 inventories, 4n10
 irrigation, 90
 mills, 88n99
 passage rights, 122
 pastoralism, 96, 106, 110–111
 political importance, 64n15
 rents, 124
 tithes and churches, 17, 67n33
Fontfroide, grange of Silvanès, 21n39, 67, 72
Forc, grange of Gimont, 45n64, 108n64
Forest, 11, 14, 25, 97, 98
Forges, 87
Foulque, bishop of Toulouse, 33n17
Foundation circumstances, 1, 5–10, 31–34, 38–39, 86, 118, 125, 127–28
Fragos, parish of, 96n7
Fraisse, *mansus* of, 20
Fraissenet, parish of St. John of, 26n57, 98
Franquevaux, abbey of,
 acquisitions, 24–25, 26, 29
 church, 4
 clearance, 26
 commercial ties, 4
 conveyances by: Anastais, 25n50; Bremund of Uzès, 111n78; Bremund William of Sommières, 25n50; Raymond V, count of Toulouse, 25n55; William Hipolytus, 25n55; William of Montpellier, 39n40
 documents, 4
 financial condition, 118
 granges, 4
 hospice, 4n12, 123
 mills, 88n99
 pastoralism and transhumance, 111
 salt, 102
 site, 4, 24
Franqueville, grange of Gimont, 45n64, 63, 108n64
Frotard of Turre, 109n68
Fulling and fulling mills, 87–89, 120, 123

G. of la Roca, 21n37
Gago of Peyrebrune, 35n26
Gaillac, grange of Silvanès, 67, 96n7
Gaillac, town of, 71, 123
Gallard of Sirac, 45n64
Gallia Christiana, 113
Garonne River:
 mills on, 89
 pasture near, 104–05, 107–08
 reclamation along, 27–28
 toll exemptions on, 82–83, 122
Garrigues, 110, 111
Gasaille contracts, 116
Gascon people, 65n16
Gascony: 16, 91–92, 104–07
Gassiarnaldus of Marrast, 13n7

INDEX

Gassias Ciche, 48n74
Gaugiac, territory of, 22, 38n36
Gausbert, abbot of Candeil, 13n5, 35n26, 76n71
Gavère, forest of, 98n17
Genciac, church of, 65n21
General Chapter, 86, 113–114, 122, 127–128
General Chapter legislation. See *Statuta*
Gerald Espaniol, 38n36
Gerald of la Paille of Mèze, 44
Gerald of Salles, 2, 32, 33, 112
Gerald of Wales, 7
Gervaissenc, *mansus* of, 50n86
Gifts: See also recruitment
    at death, or in monastic infirmary, 20, 42, 49n78, 102
    de caritative, 13n5, n6, n7, 19n29, 36, 50n86
    for foundation of a house, 36n38
    for love of God and redemption of sins, 21n37
    for soul of deceased, 42n50
    in recompense for violence, 42
    in wills, 38–39
Gimont, abbey of:
    abbots: Bernard 28n65, 47n70; William 87n95
    casaux, 16
    charters and cartulary, 2, 43n53, 66n24, 107n59
    clearance, 28
    commercial interests, 122
    conveyances by: Bonus of Perissan, 47n70; Deusaiuda of Lombirag, 47n70; Gallard of Sirac, 45n64; Sancius of Armadanvilla, 28n65; William of Sirac, 47
    direct management, 77
    disputes, 41n48, 45n64
    foundation, 2
    granges described, 66
    granges: Forc, 45n64; Franquevaux, 45n64, 63, 108n64; Hour, 48; Laurs, 16, 22,
    irrigation, 90
    mills, 87n95, 88n99
    pastoralism, 106–108
    recruiting, 78
    site, 2
    tithes and churches, 66n25
Glandage, 98
Goats, 96
Goiac, *villa* of, 22
Gondor, parish of, 65
Gorc Centones, territory of, 66n22
Gothmerlenc, forest of, 98
Grand Causses, 114–15
Grandmontines, 97n12
Grandselve, abbey of:
    abbot as arbiter, 124
    abbots: Arnald Amalric, later abbot of Cîteaux, archbishop of Narbonne, 64n15; Bernard, 87n95; Bertrand, 45n65; William, 27n62, 108n65

bastide of Grenade, 63n11
burial place, 39n40, 42n50
cartularies and other documents, 2n3, 22n40, 120n9
clearance, 20, 27–29
commercial ties, 69n44
direct management, 78
disputes, 41n48, 45n64
conveyances by: Angevin kings, 82, 102; lords of Armagnac, 107; Arnold Helye, 45–46; lords of Béarn, 107; Bernard of Montaut, 47; Bernard Pilgrim, 46; Bernard Terz, 38n36; lords of Comminges, 107; lords of Foix, 107; Gerald Espaniol, 38n36; lords of Lesparre, 102; viscounts of Narbonne, 102; Odo of Lomagne, 45n50; Peter of Bagnols, 45–46; Pons Amans, 108n65; Raymond Vidal, 57; Sasneus, 27n62; Terrenus, 38n36; counts of Toulouse, 107; Vidal of Bairis, 20; William of Montpellier, 39n40
expenditures, 38
forest rights, 98
foundation, 2–3, 32
granges described, 66n26, 68–69
granges: Bagnols, 45–46, 69n41; Bolhaguet, 69n41; Calcassac, 69n41; Comberouger, 69n41; Coubirac, 69n41; La Terride, 22, 45n64, 69n41, 108n64; Larra, St. John of, 22, 69n41, 108; Lassale, 69n41; Nonas, 69n41, 98; Rieumanent, 69n41; Salet (Lescout), 69n41; Vieilleaigue, 22n42, 69n41; Villelongue, 69n41
hermits at, 3n8, 32
hospice, 57n114, 123
mills, 87–89, 91
mortgages, 27
olive groves, 74
pasture rights and transhumance, 74, 107–110
patrons, 39n40
political importance, 64n15
recruiting, 78
reformers, ties to, 32
salt, 102
site, 2, 27
tithes and churches, 51n89
toll exemptions, 82–83, 122
urban properties, 123
vineyards and viticulture, 22n40, 24, 57n114, 92–93
Granges. See also individual abbeys.
    acreages, 63
    administrators: grangers or grange masters, 62
    advantages of, 59–61, 66, 70, 75, 93, 126
    agriculture on. See agricultural practice
    attacked in Hundred Years' War, 63
    became bastides, 120
    buildings, 61, 63, 90
    creation of: 10, 45n64, 52, 61, 61–66

direct management on. See agricultural practice
distribution and diversity of Cistercian, 40, 62–66, 69–74
elsewhere in Europe, 62
labor on, done by *conversi* and hired laborers, 56
productivity, 61, 64
replacing seigneurial estates, 128
sites, 11, 72
size, 62–66, 75
transforming countryside, 54, 64–66
transit stations, 69–70
Grausone, grange of Silvanès, 67, 72, 96n7
Gregorian reform, 52–53
Gregory IX, 33n17
Grenade, *bastide* of, 63n11
Grenoble, 4, 105
Grinding, 87, 120
Guarantors for transactions, 35n29
Guigo, count of Vienne, 105n50
Guillelm family, Montpellier, 4, 39
Guillelma, 109n68
Guillelm, abbot of Bonneval, 113n85
Guillelm Gassias of Orbezan, 42n49
Guillerma, wife of Ugo of Serruz, 19n29
Guillerm Ademari of Montealegro, 19n29
Guirald, abbot of Silvanès, 95n6, 96n7, 102n39
Guraldinus, mansus of, 25n50
Guy de Bourgogne, Calixtus II, 36
Guy Guerriatus, 39n40, 89n101

Haulterive, grange of Fontfroide, 67n33
Hauteville, territory of, 22
Hay production, 88–89, 92
Hector of Mirabel, 35n26
Hérault River, mills on, 89
Heremus, 26
Heresy, 10, 42, 124, 128–29
Hermits and hermitages, 3n8, 27, 29, 32, 34n21, 62, 97n12
Hers River, in the Ariège, 29, 110
Hers River, east of Toulouse, 108
Hired laborers, 55, 74, 79–82, 104
Horses, 40, 95, 96
Hospices, 4n12, 50n86, 57n114, 121, 123, 124
Hospitallers, 32, 41n48, 94
Hospitals, 108n63
Hospitum, 13
Hugh, abbot of Berdoues, 99n26
Hugh Francigena, author of Silvanès chronicle, 127–28
Hugh of Gaillac, 20
Hunting, 97

Improvements as index of settled land, 15
Improvements by Cistercians, 23–24, 87–90
Incorporation of earlier religious houses, 2–4, 31–33

Incultus, 14–15, 27
Indemnities, 124
Industrial uses of mills, 91, 123
Industries, wool and leather, 122
Inflation, 85
Innocent II, 50n88
Innocent III, 67
Investment in agriculture, 37, 83, 86, 90
Irrigation, 26, 27, 89–90, 120
Isarn Jourdain of Saissac, 47
Ismio Ruvoiria, 25n54
Iz, grange of Bonnecombe, 44n56, 52, 64–65, 68n36–39, 71–72

John, abbot of Bonnevaux, 105n50
Jourdain of Marrast, 13
Jourdain, *mansus* and *bordaria* of, 44–45
Judicial rights, 91

Knights and Cistercians, 34–35

La Barthe, mills at, 88n96
La Brugaireta, *mansus* of, 38n36
La Calm, *mansus* of, 38n36
La Calvarie, 111
La Capelle, 41n48
La Capelle d'Arboissel, 68n37
La Carpentaire, *mansus* of, 20
La Coste, *mansus* and *caputmansus* of, 13
La Couvertoirade, 111
La Côte Asiné, territory of, 66n22
La Foisse, or Foissa, holding of, 22
La Fraissenet, grange of Bonneval, 67
La Gard-Dieu, abbey of, 2n3, 118
La Grande Chartreuse, 105
La Graveira, *casal* of, 28n63
La Malaira, *mansus* of, 38n36
La Pardieu. See Pardieu
La Riusa, mills at, 88n96
La Rochelongue, parish of, 17
La Rode, later site of Ardorel, 73n65
La Roquette, grange of Bonneval, 67, 72, 113, 114
La Roquette, territory east of Rhône, 26
La Rovière, territory of, 25n50
La Selve, hospital of, 57n114
La Serre, grange of Bonneval, 38n36, 67, 71–72, 103, 114. See also Bonnefon
La Terride, grange of Grandselve, 22, 45n64, 69n41, 108n64
La Valle, *mansus* of, 76n71
La Vayssière, grange of Bonneval, (photo, xiv) 72
Labor costs for Cistercians, 74, 76, 77–78, 99–100
Laborers, for creation of *villeneuves*, 30, 48, 125n36
Labor-intensive activity, 93
Labor-saving, 90–92, 120–121
Lacome, *casal* of, 20

INDEX

Lady Assault, 19
Lafite, territory of, 46–47, 48n74
Lafon. See Magrin
Laity, 34–35, 88–89, 92, 94
Lambert of Castelpers, 35n26
Land acquisition:
   adjacent holdings, 28n63, 43–44
   cash expended for, 23, 37
   compacting, 46–50
   consolidation and reorganization, 12, 21–23, 31–61, 64–65, 75, 78, 93, 118–19, 125, 128
   completion, 52, 65
   contracts, types of, 36, 43–50
   cultivated lands, 11, 13, 16, 22
   descriptions of land, 12, 27n62
   expenditures for, 23, 27, 38
   fragmentation, 20–23
   improved land, 15
   making land cultivable, 27n58
   pace of, 43n54
   recently reclaimed land, 23–24, 46
   recordkeeping and planning, 41–42
   results of reorganization, 119
   ties to recruitment, 56, 125
   tithe exemption, 84
Land without any lord, 20
Landes, territory of, 56
Landola, land at, 35n25
Landrins, parish of, 16
Larra, St. John of, grange or cattle-station of Grandselve, 22, 69n41, 108
Larzac, 95, 111–113
Lavaur, forest of, 98
Las Clotas, tenants at, 57
Las Lobnatières, territory of, 66n22
Lasmeades, *casal* of, 46–47
Lassale, grange of Grandselve, 69n41
Lassouts, grange of Silvanès, 72, 77n75
Laurs, grange of Gimont, 16, 22
Laurs, vineyards of Silvanès, 35n26
Lauzig of Stela, 42n51
Lavazencs, holding of, 16n15
Lay brothers. See *conversi*
Layering of rights, 22–23
Le Thoronet, abbey of, 4
Legal settlements, 41. See also *convenientia*
Legitimization of power, 35
Lekai, Louis J., 128
Lending. See mortgages
Lenthio, grange of Léoncel, 67
Léoncel, abbey of:
   confirmations, 67
   donors: Arbert of Turre, 96n8
   financial condition, 118
   foundation, 4
   granges described, 66, 73
   granges: Cognerio, 67; Combacalida, 67; Lenthio, 67; Pallaranges, 67; Vulpa, 67
   irrigation, 26
   mills, 88n99

   pastoralism and transhumance, 67, 96, 105, 115n101, 116
   priory of Pardieu, 33, 34n22, 73
   site, 4
   tithes, 67n28, n29, 116
Les Bourines, grange of hospital of Aubrac, 63n12
Les Chambons, abbey of, 115, 123
Les Gallinières, grange of Bonneval, 62, 63n12, 72
Les Ortes, grange of Valmagne, 69
Les-Vaux-de-Cernay, abbey of, 24n49
Lesparre, lords of, 102
Lévézou, region of, 101, 115–16
Limosa, church and parish of, 52, 64n14, 65
Livières, grange of Fontfroide, 67n33
Local conditions, adaptation to, 10
Locdieu, abbey of:
   abbots: Arbertus, 113n85
   disputes, 114;
   documents, 113n82
   financial difficulties, 31, 69, 112
   foundation, 112–113
   granges, 69
   confirmations, 67
   pastoralism, 112–15
Lodève, warehouse in, 123
Lomagne, pasture in, 108n63
Lordship, 16, 20, 21, 46, 58, 71
Loubeville, territory of, 22, 22n42
Lozère, 115n97
Lubéron, pasture in, 111
Lyons, Bonnevaux's ties to, 4

Magdarz, *villa* of, 49
Magrin or Lafon, grange of Bonnecombe, 15n11, 20–21, 44n56, 68n36–39, 70
Malum Montem, territory of, 67n29
Managerial abilities and practices, 31, 59–61, 87–93, 120, 125
Manneville, territory of, 48
Mansedieu, grange of Bonnecombe, 68, 113n88
Mansus, mansi, 13, 15, 21, 49, 64, 65
Mansus al Segalairil, 71
Mansus hereminius et condrictus, 27
Manual labor, importance to order, 41, 126–127
Marcillac, *vallon* of, 71
Marginal land, 12, 23, 76–77
Margnes, grange of Silvanès, 67, 72, 78, 111
Markets and market privileges, 96–97, 100, 121–24, 129
Marquisate of Provence, see Provence, Marquisate of
Marshlands. See *nausa, paluda*, salt-marshes
Martres, *mansus* of, 13
Masdoat, territory of, 22
Maselz, *mansus* and *caputmansus* of, 20, 21n37

Masse, grange of Bonneval, 67, 72
Massif Central, pasture on, 105, 110–115
Mated, forest of, 98
Matfres, 76n71
Matilda, countess of Vienne, 105n50
Mazan, abbey of, 112–115, 123
Mazeriis, *condamina* of, 45n65, 46n66
Meadows, 49, 76–77, 88–89
Medium crescementum, 115–116
Medium fenum, 116n104
Medium vestum, 87n95
Mendicant orders, 129
Mercurières, grange of Valmagne, 44, 66, 69, 73
Mèze, town of, 122
Middlemen, 22–23, 77
Millau, 68n38, 114
Mills, 24, 27n58, 49, 87–91, 120, 123
Millstones, 88n96
Minhac, grange of Bonnefont, 70, 72
Mirabel, territory of, 68n38, 101
Mocenes, *mansus* of, 57n114
Moissac, abbey of, 28n64, 39, 41n48, 120n9
Moltura, 87
Monastic patrimony. See granges
Moncan, grange of Bonnecombe, 68n36–39, 70–71
Money, 37–38
Mont Ventalos, territory of, 66n22
Montagne Noire, 73, 111
Montagnes de Ravat, 109
Montagnes, 95, 109–111. See also pastoralism and transhumance—summer pastures.
Montagne de Muson, 67n29
Montarsin of Calmont, 35n26, 39n41
Montégut, grange of Bonneval, 67
Montesquieu family, 37n30
Montlaurès, church and grange of Fontfroide, 17, 67n33
Montpellier, commercial ties to, 72
Montpellier, Guillelm family, 4, 39
Montpellier, hospice, 122–23
Montpré, land at, 47
Montrédon, grange of Fontfroide, 67n33
Montveyre, church and monastery of, 17, 67n33
Moral advantage of monks over neighbors, 42
Morimond, abbey of:
  abbot Wualterius, 36n28
  filiation of, 2, 69, 106
Morlencs, mill at, 88n96
Mortgages, 16n15, 23n45, 37, 39, 99n27, 112–113
Motives for foundation, 34–35
Mousnier, Mireille, 68
Moving of animals. See pastoralism and transhumance.
Murder, 58–59

Nadesse River, mills at its confluence with Garonne, 89n101

Narbonne, archbishop of, 102; Arnald Amalric, 64n15
Narbonne, commercial ties to, 122
Narbonne, viscounts of, 102
Narrative descriptions of Cistercian foundations, 6, 9, 127–28
Naucelle, vicinity of, 68n38, 71
Nausa (marsh), 27n62, 28n63
Net yields increased, 75, 79–82, 84, 144–47
New land, 12, 15, 28, 30, 86
New towns, 120. See *bastides, sauvetés, villeneuves*
Nîmes, 4, 123
Nizors, abbey of, 110
Nonas, grange of Grandselve, 69n41, 98
Nonenque, abbey of women, pasture rights of, 111–12
Notables, 34–35
Noval lands, 12, 15, 28, 86
Noval tithes, 15n11, 18, 18n23, 52, 86n93
Nuns, x, 24n49, 94, 108n63, 111–12

Odo of Graviano, 57n114
Odo of Lomagne, viscount, 42n50
Olive production, 24, 70, 72
Omellaz, territory of, 96
Oms, territory of, 14
Onerous conveyances, 36
Oneth, *villa* of, 44n55, 64n13, 68n38
Otencs, land of, 68n37–38
Ovens, 90–91
Ozalth, forest of, 98

P. abbot of Bonnecombe, 91n111
Pagesia (rights to cultivate), 18–19, 21, 55–56
Pallaranges, grange of Léoncel, 67
Paluda (marsh), 25n55
Panat family, 35n26, 37n30
Paollan. See Paulhan
Papagall, *caputmansus* of, 21n37
Papal confirmations, 26, 50n88, 66–68, 83, 86. See also individual abbeys
Pardieu, priory of, 33, 34, 73
Pardinéguas, grange of Silvanès, 67
Pardines, grange of Fontfroide, 67n33
Pardiniac, territory of, 44, 66n22
Parish priest conveyed tithes, 51
Parishes and parish churches, 16–17, 65
Parlant, *appendaria* of, 15n11, 44n56
Passage tolls, exemption from, 40, 99–100, 102, 107
Passar, apojar et adavallar, 95n3
Pastoral imagery, 97n12
Pastoralism and transhumance, 70, 73, 84, 92, 94–117, 126. See also individual abbeys
  acquisition of pasture, 94, 98–99
  advantages of, 94, 99–100, 115, 117
  aliena animalia (outsiders' animals), 115–16
  animal products, 96
  ascent and descent of animals, 95

INDEX

breeding, 90, 100, 117
cattle, 95, 96, 103
commercialization of, 114–15
contracts, 115–116. See also contracts
cooperation in, 94
demand for animal products, 96
disputes, 95, 101, 103, 107–08, 109–15
documents, 94–95, 101–17
donors of, 95n2, 98–99, 100n32, 107
extensiveness of, 94–95, 101–04, 115
eremitical groups and, 97n12
forest pasture, 95, 97–98
gasaille contracts, 116
glandage, 98
goats, 96
horses, 40, 95, 96
influenced practice of neighbors, 88–89, 92, 94
isolation, order's reputation for, 97
locations: Alps, 4, 67, 105–06; Cevennes, 73; Durance and Isere River valleys, 4; Gimont's grange of Laurs, 16; between Aveyron and Viaur Rivers, 113–114; between Tarn and Garonne Rivers, 108n65; Rouergue, 101
locating, 95
limits on order's use, 101, 108n65
market for pasture rights, 100–101
markets for animals, 117
maximization of pasture usage, 103–104
meadows and hay-making, 49, 76–77, 88–89, 99n26
monopolized by stronger abbeys, 112–115
motivations for development, 96–97
numbers of animals, 95–96, 99, 116
passage rights, exemption from tolls, 40, 99–100, 102, 107
pigs, 95–98
pioneering, order's reputation for, 97
purchase of animals, 86
ratio of pasture to arable land, 99
routes and roads, 95, 105, 114
salt, 102
sheep, 95–96, 100
shepherds, 104
shepherds' huts, 98, 109
start-up costs, 99–100
summer pasture, 95, 99, 104, 106, 109–10, 116
success of order in, 38–40
tithes on, 100, 116
wasteland and fallow, 98–99
water, *abreuvage*, 109n68
Patrons, 34
Paulette, grange of Bonneval, 67
Paulhan or Paollan, mills of, 39n40, 89
Paunac, grange of Valmagne, 69
Peace of Paris, 34n24, 124
Peasants. See also dependent-tenants, *pagesia*
as proprietors, 18–20.
became *conversi*, 19, 35, 47, 49, 56. See also *conversi*, and recruitment
forbidden to use Cistercian mills, 87
on Cistercian land, removal of, 20, 48, 53–59
who did not become *conversi*, 47, 57–59
undertook reclamation, 47–48, 125n36
Peirola, forest of, 98
Perambulations, 44, 47
Peritio, *mansus* and *caputmansus*, 19n29
Permission to clear, 24–29
Personnel. See recruitment
Peter, bishop of Rodez, 65n21
Peter Abole, 44
Peter Carpenter, 20
Peter Durand of Bezins, 44
Peter of Bagnols, 45–46
Peter of Castelnau, 53n15
Peter of Caucill, 35n25
Peter of Orbezan, 42n49
Peter of Puy Laurent, 77n75
Peter of Teillz, 21n37
Peter of Turre, 35n26
Peter Rainaldi, 56n110
Peter Ugo, 76n71
Peter Ymbert, 35n26, 38n35
Petronilla, wife of Bernard of la Roca, 21n37
Peyrebrune, castle of, 101n35
Philip the Fair, 113n82
Philippa of Toulouse, 32
Phraseology of charters, 12–16
Pigs, 95–98
Pilgrimage, 39
Pin, mill at, 89n101
Pleus, parish of, 65
Poblet, abbey of, 96
Pojet, *mansus* of, 38n36
Political power, 112–13, 128–29
Pomers, grange of Bonneval, 67
Poncia, sister of Arnaldus of Taurinus, 13n5
Pons, abbot of Silvanès, 19n29, 77n75, 102n39, 127n39
Pons, abbot of Bonnecombe, 13n6
Pons Amans, 108n65
Pons Bertold, 44
Ponsan, men of, 59, 59n120
Pontigny, filiation of, 69, 112
Poujouls, grange of Fontfroide, 67n33
Pousthoumy, 68n38
Poverty and asceticism. See Cistercian order—asceticism
Pradela, land at, 28n63, 38n36
Praemonstratensian canons, controversy over pasture, 94
Pre-Cistercians, 26–30, 33
Preferred status sought, 121
Previnquières, grange of Bonneval, 67
Price of land, 23, 37
Priories, incorporated, 34
Privileged status, 40–41, 121, 129
Production geared to markets, 76
Profits and profitability, 23, 40, 61, 86, 93, 100

Promillac, grange of Silvanès, 19n29, 22, 27n61, 49, 67, 72, 87n94, 88–89
Propaganda, 8, 129
Provence, Cecilia countess of, 4
Provence, marquisate of, 4, 16, 96, 105–106
Prugnes, parish of, 52
Puechmaynade, land of, 71
Puszac, grange of Bonneval, 67, 72
Puy d'Oneth, *mansus* of, 38n35
Puy, *mansus* of, 13
Pyrenees, 70, 73, 105, 106–110

Quantitative measures of earlier settlement, 15–16
Quartum, 13, 80n79
Quercy, transhumance, 115

Rabastens, forest of, 98
Randulf, abbot of Bonnecombe, 19n27
Ravat, summer pastures of, 109–10
Raymond, count of Toulouse. See Toulouse, count of
Raymond, count of Barcelona, 82
Raymond Carbonelli, cellarer of Bonnecombe, 38n36
Raymond Guillermi of Fabrezano, 77n75
Raymond Hospitale, 48n74
Raymond of Castronovo, 35n26
Raymond of St. Germain, 19
Raymond Richard of Foderia, 95n6
Raymond Sanche of Cortade, 34n23
Raymond Trencavel, viscount of Béziers, 34n24, 82
Raymond Vidal, 57
Reassembling, see consolidation
Reclamation, 11–12, 27–29
Reconstitution of ownership. See consolidation
Recruitment, 13, 34, 35n25, 37, 55–57, 78, 125, 126
Reform groups, incorporated or annexed, 31–34
Reformers of western France, 1, 32–33. See Robert of Arbrissel, Gerald of Salles
Regina, 93n119
Relationships within order. See filiation
Religious aspirations, 35
Religious benefits for donors, 13, 34, 37, 38n36, 42n50. See also gifts, recruitment
Religious services restored at parish church, 48–49
Rents owned, 76n71, 82–83
Repopulation by creation of *bastides*, 120
Restapauc, *mansus* of, 38n36
Revolutionary government, sale of monastic holdings, 63
Rhône River, 4, 104–105, 114, 122
Ricancelle, church of St. Sernin of, 51n89
Rieumanent, grange of Grandselve, 69n41

Rigaldus Amblartz, 21n37
Rights in all my lands, 16
Rignac, pasture near, 101
Riparia, *mansus* of, 13
Risk, spreading of, 70, 74
Robert of Arbrissel, 32
Rodez, bishop and canons of, 38n36, 49, 51
Rodez, commercial rights in, 122–123
Rodez, counts of, 23n44, 35n29, 37n30
Rodez, houses in, 68n37, 123
Roger, nephew of Peter of Serres, 42n50
Romanized areas, 44
Rotation of crops, 91–92
Roterius of Solis, 35n26
Rouergue, 15, 104n44, 105, 112–115
Rouzet, grange of Silvanès, 56n110, 67, 72
Rover, *mansus* of, 44n56
Ruffepeyre, holding of, 71
Rupta, clearance, uprooting of trees, 25n54

Salavert, grange of Fontfroide, 67n33
Sale disguised as donation, 36n29. See also contracts
Salet (Lescout), grange of Grandselve, 69n41
Salinarias. See salt-production
Salt and salt production, 26, 27, 82, 102
Salt-marshes, 24–25, 26
Sanche Lub of Faisan, 42n50
Sanche, son of Bernard of Astarac, 36n28
Sanche of Armadanvilla, 28n65
Sanche of Comères, 34n23
Sasneus, wife of late Cenabrus of Canzas, 27n62
Saunnac, *mansus* at, 57n114
Sauvelade, abbey of: incorporation, 118
Sauveplane, grange of Silvanès, 67, 72, 78
Sauveréal, abbey of: 24, 82, 111
Sauvetés, 46n68, 120n9
Sauze, forest of, 98n19
Savardun, *montagnes* of, 110
Save River, 108n63
Savings and frugality, 41, 75, 81–83
Seed, 90
Seed/yield ratios, See yields
Sénanque, abbey of, 4, 119n5
Sénones, or Cénomes, 65n21
Serreneuve, forest of, 98
Serres, castle of, 14, 57n114
Serruz, parish and church of, 49, 56, 65n21, 96, 111n79
Services owned by tenants to Cistercians, 19, 47, 57
Settled land, 11–25, 27, 45
Sevalz, vines at, 57n114
Séveyrac, grange of Bonneval, 72
Sheep roads (*drailles, drayas*), 105, 114
Sheep, 95–96, 100
Shepherds' huts (*cabanes*), 98, 109
Shepherds, 104
Sicard Alaman, owner of tolls in region of Albi, 107n62

INDEX 177

Silva, replaced by *villeneuve*, 48
Silva Arriano, Gimont built grange there, 45n64
Silva Godesca, pasture at, 111n78
Silvacane, 4
Silvanès, abbey of:
  abbots: Guirald, 95n6, 7, 102n39; Pons, 19n29, 77n75, 95n6, 102n39
  cartulary, 3, 43n54
  chronicle, 3, 38n39, 39, 86n92, 127–128
  church, 119n5
  churches and tithes, 49, 65n21
  competition, 73, 86n92, 127
  consolidation, 15, 21–22
  conveyances by: Berengar Rotbald of Capéstang, 102; Berengar, archbishop of Narbonne, 102n39; Bertrand de Ponte and brothers, 96n7; bishops of Rodez, 49; Englesa, 77n75; Guillerma, wife of Ugo of Serruz, 19n29; Guillermus Ademari of Montealegro, 19n29; Jordan of Foderia, 95n6; archbishop of Narbonne, 102; Peter of Puy Laurent, 77n75; Raimund Guilermi of Fabrezano, 77n75; Raymond Richard of Foderia, 95n6; Ugo of Serruz, 19n29
  demographic increase near, 48n74, 125n36
  direct management, 78
  disputes, 3, 41n48
  granges described, 72
  granges: Fontfroide, 21n39, 67, 72; Gaillac, 67; Grausone, 67, 72, 96n7; Lassouts, 72, 77n75; Margnes, 67, 72, 78, 111; Pardineguas, 67; Promillac, 19n29, 22, 27n61, 49, 67, 72, 87n94, 88–89; Rouzet, 67; Sauveplane, 67, 72, 78; Sollies, 67, 111n79
  hermits there, 17, 32
  hospice, 122–123
  irrigation, 27, 90
  mills, 49, 87n94, 88–89
  origins, 3, 39, 113
  papal confirmations, 67
  pastoralism and transhumance, 77n75, 95–96, 111–113
  received *mansus hereminius*, 27
  recruiting, 35n25, 56, 78
  salt, 102
  site, 27
  tithes, 52, 65n21
  vineyards, 36n26
Simon of Elbone, 116n104
Simon of Montfort, 124
Simony, 51
Sisteron, *montagnes* of, 111
Size of granges, 60
Size of southern French Cistercian houses, 118
Smaller abbeys, 94, 112–15
Solbario, territory of, 22
Sollies, grange of Silvanès, 67, 111n79

Spain, 4, 96, 106
St. Amans of Sénones, 65n21
St. Anthony of Pamiers, abbey of, 33n18
St. Benedict, *silva* of, 47
St. Cyprien, olive cultivation there, 74
St. Egez, vines at, 68n36–39
St. Elix, grange of Berdoues, 62n4
St. Felix of Valence, 116
St. Félix, grange of Bonnecombe, 62n4, 64, 68n38–39, 71, 72
St. Gaudens, 109
St. George, parish of, 22
St. Gilles, town of, 24, 115, 122
St. Jacob, route of, 108n63
St. James of Villevalerian, territory of, 47
St. John of Villelongue, *villa* of. See Villelongue
St. John of Valségier, *villa* of, 47
St. John of Larra, grange of. See Larra
St. John of Fraissenet, parish of, 26n57, 98
St. John of Celsan, parish of, 17
St. Jory, mills at, 87n94
St. Julian, cellar of, 67n28
St. Just of Durban, parish of, and grange of Fontfroide, 17, 67n33
St. Lawrence, *nausa* of, 27n62
St. Marcel, abbey of, 2n3
St. Martin de la Garrigue. See Burau
St. Martin le Vieux, territory of, 47
St. Martin of Toget, parish of, 48
St. Martin of Tels, territory of, 98
St. Martin of Colons, parish of, 17
St. Martin, grange of Fontfroide, 67n33
St. Martin, parish of, 47
St. Mary of Varnassonne, territory of, 47
St. Maurice, parish of, 16
St. Michael of Cuxa, abbey of, 33
St. Paul, grange of Valmagne, 69
St. Peter de la Bourg de Valence, 116
St. Peter of Beceria, parish of, 17
St. Peter of Vinsac, parish of, 22
St. Ruf, church of, 20
St. Sernin of Ricancelle, church of, 51n89
St. Sernin, abbey of, 45–46, 120n9
St. Sernin, parish of, 105n50
St. Severin, church of, 22n42
St. Stephen of Durfort, church of. See Durfort
St. Sulpice of Boulhac, church of, 51n89
St. Victor of Montveyre, 17, 67n33
St. Ylarius, gardens and property at, 38n36, 68nn37, 38
Staggering of agricultural work, 74
Stagnum, pond or salt-pond, 27n58
Statuta, 5–6, 9n27, 53, 54n102, 69, 86n91, 87n95, 104n44, 113–14, 115n97, 125n35, 127
Ste. Croix de Serruz. See Serruz
Ste. Eugenie, grange of Fontfroide, 67n33
Ste. Eulalie, hospital of, pasture rights, 111–15
Stock-raising. See pastoralism

Stone-cutting, 87
Subsistence, 81
Successful Cistercian houses, 33
Summer heat, 73
Surnames used to locate pasture rights, 95
Sylvo-pastoralism, 97–98

Tanning, 123
Tarascon-sur-Ariège, 109, 110
Tarn River, 107–08, 114
Taura, grange of Fontfroide, 67n33
Taurinus, *stagnum* of, 27n58
Tax rolls of order, 118
Tegula, grange of Bonneval, 67
Tels, territory of St. Martin of, 98
Templars, 41n48, 94, 111–113
Tenants. See peasants, dependent-tenants
Tenementum, Tenementa, 15, 65
Tenurial rights conveyed, 55
Terminology, 12–16, 29–30
Terra culta et inculta, 14–15, 27
Terundum, *mansus* of, 27n60
Teuleiras, *mansus* of, 38n36
Theft, 42n51
Tirapel, territory of, 22
Tithes, ecclesiastical:
  and churches, 15–18, 50–53
  collection, 84
  disputes, 51–52, 115
  exemption, advantages of, 83–86; animals included, 13, 52, 100, 115–16; complaints, 86; efforts to retain, 86; first fruits included, 83; limits, 52, 86
  fragmented, 21
  noval tithes, 15n11, 18, 52, 86n93
  granted upon departure for Crusade, 35n26
  repurchase, 46, 50–53, 83–85
Toalis, *villa* of, 20
Toeils, castle of, 101n35
Toget, castle of, 101n35
Tolls and market taxes, 75, 82, 95, 122
Tortoreria, *stagnum* of, 27n58
Toubert, Pierre, 11
Toulouse, bishop of, 3n8, 32, 51n89: 33n17, 129n42
Toulouse, city of, 57n114, 70, 105, 107–08, 114, 122–23
Toulouse, count of, 23n44, 25n55, 28n64, 34, 39n38, n39, 82, 91n111, 107,124
Towns:
  growth of demand in, 96–97
  properties in, 121–23
  ties to, 117, 118, 121–24, 126, 129
Tramesaigues, priory of and grange of Boulbonne, 33, 34n21, 73n65
Transformation of countryside, 128–29
Transhumance. See pastoralism and transhumance, 109
Treasure in circulation, 120
Tremoillae, church of, 35n29

Trencavel family. See Cecilia of Provence, Raymond Trencavel
Tresmanses, grange of Valmagne, 69

Ugo, abbot of St. Sernin, 45n65
Ugo, abbot of Bonnecombe, 21n37
Ugo, count of Rodez, 21n37, 35n29
Ugo Bonefons, 35n25
Ugo of Durfort, 42n50
Ugo of la Roca, 35n26
Ugo of Salvinniac, 35n26
Ugo of Serruz, 19n29
Ugo of Seveirac, 36n29
Ugonencs, *mansus* of, 20, 50n86
Ulmet, abbey of. See Sauveréal
Uncultivated land, 11, 14–15, 27
Underemployment, 74n68
Unencumbered lands, 17
Unprofitable holdings, 85
Untitled men, 35n25
Upheaval caused by Cistercians, 119
Usac, territory of, 22
Usatica, 19

Vairac, grange of Valmagne, 25n56, 27n58, 38, 48, 69, 73
Vajal, priory of and grange of Boulbonne, 29n66, 33, 34n21
Val Lutosam, 67n29
Valautre, grange of Valmagne, 39n40, 69, 73
Valcroissant, abbey of: 98n17, 105
Valence, churches in, 116
Valle Paillède, territory of, 66n22
Valmagne, abbey of:
  assarts acquired, 29
  cartulary, 4n11, 34n24, 69
  churches and parishes, 16–17
  commercial interests, 4, 69n44, 122
  competition, 73, 86n92, 127
  confirmations, 26
  consolidation, 48
  conveyances by: Aimeric of St. Paragoire, 44; Alphonse, king of Aragon, 82; Bernardus Ermengaudi, with brothers Raymond and Gaucelmus, 27n58; Cecilia of Provence, 4; Gerald of la Paille of Mèze, 44; Guy Guerregiatus, 39n40; Peter Abole, 44; Peter Durand of Bézins, 44; Pon Bertold, 44; Raymond Trencavel, viscount of Béziers, 82; Raymond, count of Barcelona, 82; William of Montpellier, 39n40
  donations, 124n31
  draining marsh to create milling complex, 26
  exemption from tolls, 82
  expenditures, 38
  fishing rights, 4, 82, 122
  foundation and incorporation, 4, 32, 34n24

forest rights, 73, 98
granges, described, 4, 66n26, 69
granges: Burau, 26, 66n17, 69, 73; Canvern, 69, 73; Capriliis, 69; Champdesvignes, 69; Fondouce, 69; les Ortes, 69; Mercurières, 44, 66, 69, 73; Paunac, 69; St. Paul, 69; Tresmanses, 69; Vairac, 25n56, 27n58, 38, 48, 69, 73; Valautre, 39n40, 69, 73
hospice, 122–123
mills, 88n99, 89
pastoralism and transhumance, 96, 111
salt-production, 26–27, 102
viticulture, 92–93
Valségier, *villa* of St. John of, 47
Vareilles, grange of Bonnecombe, 13, 56, 68n36–39, 70
Vernassonne, territory of St. Mary of, 47
Velocity of exchange, 120
Verned, *nausa* of, 27n62
Verruca, grange of Bonneval, 67
Viaur River, 21, 70, 101n37, 113–114
Vicaria, vicarial rights, 13
Vicious cycle of low agricultural yields, 92, 94, 99
Vidal, abbot of Berdoues, 108n63
Vidal Destipuei, 34n23
Vidal Grill of Serris, 57n114
Vidal of Bairis, 20
Vidal of Maceriis, 34n23
Vieilleaigue, grange of Grandselve, 22n42, 69n41
Vienne, archbishop of, Guy de Bourgogne, later Calixtus II, 36
Vienne, count of, 105n50
Villa, villae, 14, 16, 62
Villelongue, abbey of
  clearance, 27
  consolidation, 48
  donors: Isarn Jourdain of Saissac, 47; Regina, 93n119; Ugo of Durfort, 42n50
  forest rights, 98
  granges, 34n21, 72–73
  name change, 47
  origins, 32

pastoralism, 111
peasants, 58, 125n36
size of holdings, 65
villa of (abbey-site), 47, 65, 73
woods near, 125
Villelongue, grange of Grandselve, 69n41
Villeneuves, 25n55, 26, 48n74, 58, 125n36
Violence, murder, 42
Visitation by abbots, 106
Viticulture, 35n26, 70–71, 92–93, 125
Vivarais, 113, 115
Vulpa, grange of Léoncel, 67

Wars of Religion, 2, 31
Wastelands, 14, 97–98
Water-power, 120–121
Wealth and political power, 31, 69, 126, 128–29
Wilderness, lacking, 30
William, abbot of Grandselve, 27n62, 108n65
William, abbot of Gimont, 87n95
William, abbot of Bonnecombe, 95n2
William, abbot of Berdoues, 14n8, 20n33, 57n114
William, archbishop of Auch, 42n49
William Foramont, 38n36
William Hipolytus, 26n55
William of Laseran, 46
William of Montpellier, 35n26, 39n40
William of Sirac, 47
William Sanz de Lisos, 48n74
William Seinnorellz, 50n86
Wine. See viticulture
Wood for construction, tools, repairs, 97
Wool and leather industries, 122, 123
Work as prayer, 82
Workshops, 123
Wualterius, abbot of Morimond, 36n28

Yields, assumed, 80–81, 84–85
Yields, improvements in, 64
Yssens, mills of Bonnecombe, 68n38

Zatgairenca, wood at, 35n26

www.ingramcontent.com/pod-product-compliance
Lightning Source LLC
Chambersburg PA
CBHW080925100426
42812CB00007B/2375